Advanced Controllers Design for Hybrid Active Power Filters

Chi-Seng Lam · Cheng Gong · Wai-Kit Sou

Advanced Controllers Design for Hybrid Active Power Filters

Springer

Chi-Seng Lam
State Key Laboratory of Analog
and Mixed-Signal VLSI and Institute
of Microelectronics, Department
of Electrical and Computer Engineering,
Faculty of Science and Technology
University of Macau
Macau, China

Cheng Gong
State Key Laboratory of Analog
and Mixed-Signal VLSI and Institute
of Microelectronics
University of Macau
Macau, China

Wai-Kit Sou
State Key Laboratory of Analog
and Mixed-Signal VLSI and Institute
of Microelectronics
University of Macau
Macau, China

ISBN 978-981-97-6846-2 ISBN 978-981-97-6847-9 (eBook)
https://doi.org/10.1007/978-981-97-6847-9

This Springer imprint is published by the registered company Springer Nature Singapore Pte Ltd.
The registered company address is: 152 Beach Road, #21-01/04 Gateway East, Singapore 189721, Singapore

Preface

The field of power quality has gained increasing attention in recent years, driven by the rapid proliferation of sensitive electronic devices and the growing complexity of electrical power systems. Ensuring high power quality is essential for the reliable and efficient operation of modern electrical infrastructures. This book delves into the intricate aspects of power quality, with a particular focus on hybrid active power filters (HAPFs) and their advanced control strategies. It aims to provide a comprehensive guide for researchers, engineers, and practitioners engaged in the design, development, and implementation of HAPFs.

In this book, Chap. 1 introduces the fundamental power quality issues, the development of HAPFs, advancements in control strategies, research challenges, and the overall organization of the book. Chapters 2–5 cover various advanced control strategies for *LC*-coupling hybrid active power filter (*LC*-HAPF), starting with the design and implementation of deadbeat current control (DBCC), including mathematical modeling, stability analysis, and validation through simulation and experiments in Chap. 2. This is followed by an improved model predictive control (MPC) strategy in Chap. 3, vector proportional integral control (VPIC) in Chap. 4, and second-order sliding model control (SOSMC) in Chap. 5, each discussing respective control objectives, stability considerations, performance validations, and robustness under different conditions. Chapters 6–10 delve into various advanced control strategies for thyristor-controlled *LC*-coupling hybrid active power filter (TC*LC*-HAPF), with Chap. 6 examining the multi-quasi-proportional-resonant control (MQPRC) strategy, including resonant frequency design, proportional gain, and phase compensation, while Chap. 7 presents the H_∞ control strategy, encompassing harmonic state space modeling and optimal controller design. Chapter 8 focuses on finite-set model predictive control (FS-MPC), detailing predictive model construction, delay compensation, and performance evaluation, and Chap. 9 explores hybrid finite-set model predictive control (FS-HMPC), combining FS-MPC and hysteresis sliding mode control (SMC) for enhancing performance, culminating in Chap. 10, which introduces the second-order sliding mode observer (SOSMO) based second-order sliding mode control

(SOSMC), emphasizing observer design and digital implementation. Chapter 11 summarizes the key findings of the book and outlines prospective areas for future research in the field of power quality and HAPFs.

Macau, China

Chi-Seng Lam
Cheng Gong
Wai-Kit Sou

Acknowledgments

The research works presented in this book were carried out in the State Key Laboratory of Analog and Mixed-Signal VLSI (SKL-AMSV) and Institute of Microelectronics (IME), University of Macau (UM), Macau, China, and in Electrical Machines and Drives Laboratory, Department of Electrical and Computer Engineering (ECE), Faculty of Science and Technology (FST), UM, Macau, China. We would like to express our gratitude to The Science and Technology Development Fund, Macau SAR (FDCT) under Grant 0027/2021/A1, Grant 025/2017/A1, Grant 0028/2020/A1, Grant SKL-AMSV(UM)-2020-2022, and Grant 004/2023/SKL, and UM under Grant MYRG2017-00090-AMSV and Grant MYRG2018-00039-FST.

We would like to express our hearty and profound gratitude to Prof. Rui P. Martins (UM's Vice-Rector (Global Affairs), Founding Director of SKL-AMSV, Director of IME, and Chair Professor of FST-ECE), who built this great research platform and created an excellent research environment for us in UM. We would like to thank for his precious advice, experience sharing, encouragement, and great support.

Chi-Seng Lam would also like to express his hearty and profound gratitude to his Ph.D. supervisor Prof. Ying-Duo Han (Academician of the Chinese Academy of Engineering and Professor of Tsinghua University, China), for having opened his vision. Chi-Seng Lam appreciates him for this immensely inspirational guidance and precious advice.

Last, but certainly not least, we would like to send our heartfelt appreciation to our families, who endured our dedication to this book.

Contents

1 Introduction 1
1.1 Power Quality Issues 1
1.1.1 Understanding Power Quality Issues 1
1.1.2 Causes of Power Quality Issues 2
1.1.3 Impacts of Power Quality Issues 3
1.1.4 Importance of Current Quality Compensation 4
1.1.5 Power Quality Standards 5
1.2 Development of Current Quality Compensators 6
1.2.1 Circuit Configuration of *LC*-HAPF 8
1.2.2 Circuit Configuration of TC*LC*-HAPF 11
1.3 Development of Control Strategies for HAPFs 12
1.3.1 Control Strategies for *LC*-HAPF 12
1.3.2 Control Strategies for TC*LC*-HAPF 13
1.4 Book Organization 15
References 17

2 Deadbeat Current Control for *LC*-HAPF 21
2.1 Introduction 21
2.2 Control Concept 22
2.3 Design of Deadbeat Current Control of *LC*-HAPF 23
2.3.1 Mathematical Modeling 23
2.3.2 Stability and Parameter Design 24
2.4 Overall Control Block Diagram 26
2.5 Simulation Results 28
2.6 Experimental Results 32
2.7 Summary 36
References 36

3 Improved Model Predictive Control for *LC*-HAPF 39
3.1 Introduction 39
3.2 Design of Improved Model Predictive Control for *LC*-HAPF 40

3.2.1 Control Objective 40
3.2.2 Conventional Model Predictive Control for *LC*-HAPF 40
3.2.3 Proposed Model Predictive Control for *LC*-HAPF 42
3.2.4 Stability and Parameter Design of Model Predictive Control 43
3.3 Overall Control Block Diagram 46
3.4 Simulation Results 47
3.5 Experimental Results 52
3.6 Summary 59
References 60

4 Vector Proportional Integral Control for *LC*-HAPF Under Minimum DC-Link Voltage 61
4.1 Introduction 61
4.2 System Modeling and Minimum DC-Link Voltage Design of *LC*-HAPF 62
4.2.1 System Modeling of *LC*-HAPF 62
4.2.2 Minimum DC-Link Voltage Design of *LC*-HAPF 63
4.3 Vector Proportional Integral Controller Design of *LC*-HAPF 64
4.3.1 Derivation of VPIC 64
4.3.2 Optimal Parameter Design of VPIC 65
4.3.3 Robustness Analysis of VPIC 67
4.3.4 Digital Implementation of VPIC 69
4.4 Overall Control Block Diagram 70
4.5 Theoretical Comparative Study Among Different Controllers for *LC*-HAPF 73
4.5.1 Controllability of Current Controller with Minimum DC-Link Voltage 73
4.5.2 Closed-Loop Response Analysis of Different Current Controllers 74
4.6 Simulation Results 78
4.7 Experimental Results 80
4.8 Summary 85
References 86

5 Second Order Sliding Model Control for *LC*-HAPF 89
5.1 Introduction 89
5.2 Control Objective and Super-Twisting Controller 90
5.2.1 Control Objective 90
5.2.2 Super-Twisting Controller 91
5.3 Design of Second Order Sliding Model Control for *LC*-HAPF 93
5.3.1 Under System Stability Study 93
5.3.2 Under System Parameter Variation Consideration 96
5.3.3 Under Distorted Grid Voltage Consideration 96

5.3.4 Under Grid Frequency Variation Consideration 96
5.3.5 Final M_x Design Under Parameter Variation, Distorted Grid Voltage and Grid Frequency Variation Consideration 97
5.4 Overall Control Block Diagram 97
5.5 Simulation Results 98
5.6 Experimental Results 100
5.7 Summary 106
References 106

6 Multi-Quasi-Proportional-Resonant Control for TC*LC*-HAPF 109
6.1 Introduction 109
6.2 Modeling of TC*LC*-HAPF 110
6.2.1 Modeling of TC*LC* Part 110
6.2.2 Modeling of Current Control Loop 111
6.3 Design of Multi-Quasi-Proportional-Resonant Control for TC*LC*-HAPF 112
6.3.1 Design of the Resonant Frequency ω_n and Bandwidth ω_c 114
6.3.2 Design of the Proportional Gain K_p 114
6.3.3 Design of the Phase Compensation Angle ϕ_n 115
6.3.4 Design of the Resonant Gain K_{rn} 116
6.4 Overall Control Block Diagram 118
6.5 Simulation Results 120
6.6 Experimental Results 123
6.7 Summary 126
References 127

7 H_∞ Control for TC*LC*-HAPF 129
7.1 Introduction 129
7.2 System Modeling and Harmonic State Space Modeling of TC*LC*-HAPF 130
7.3 H∞ Optimal Controller Design for TC*LC*-HAPF 132
7.4 Overall Control Block Diagram 135
7.4.1 Simulation Results 135
7.5 Experimental Results 138
7.6 Summary 139
References 140

8 Finite-Set Model Predictive Control for TC*LC*-HAPF 143
8.1 Introduction 143
8.2 Modeling of TC*LC*-HAPF 144
8.3 Design of Finite-Set Model Predictive Control for TC*LC*-HAPF 144
8.3.1 Predictive Model Construction 144
8.3.2 Coupling Capacitor Voltage Estimation 146

8.3.3 Delay Compensation ... 148
8.3.4 Cost Function and Optimal Vector Selection ... 149
8.4 Overall Control Block Diagram ... 150
8.5 Simulation Results ... 150
8.6 Experimental Results ... 155
8.7 Summary ... 158
References ... 158

9 Finite-Set Hybrid Model Predictive Control for TC*LC*-HAPF ... 161
9.1 Introduction ... 161
9.2 Modeling of TC*LC*-HAPF ... 161
9.3 Design of Hybrid Finite-Set Model Predictive Control for TC*LC*-HAPF ... 162
9.3.1 Fundamentals of FS-MPC and Hysteresis SMC ... 162
9.3.2 Predictive Sliding Surface Function ... 163
9.3.3 Cost Function Design ... 164
9.3.4 Stability and Robustness Analysis ... 164
9.3.5 Sliding Coefficient Design ... 166
9.4 Overall Control Block Diagram ... 167
9.5 Simulation Results ... 168
9.6 Experimental Results ... 171
9.7 Summary ... 174
References ... 174

10 Second Order Sliding Model Observer Based Second Order Sliding Model Control for TC*LC*-HAPF ... 177
10.1 Introduction ... 177
10.2 Second-Order Sliding Mode Observer Design for TC*LC*-HAPF ... 178
10.3 Observer Based Second-Order Sliding Mode Controller Design for TC*LC*-HAPF ... 180
10.4 Digital Implementation of SOSMO Based SOSMC ... 182
10.5 Overall Control Block Diagram ... 184
10.6 Simulation Results ... 187
10.7 Experimental Results ... 189
10.8 Summary ... 196
References ... 198

11 Conclusions and Prospective for Further Work ... 201
11.1 Conclusions ... 201
11.2 Perspectives for Future Works ... 202

About the Authors

Chi-Seng Lam received the Ph.D. degree in electrical and electronics engineering from the University of Macau (UM), Macau, China, in 2012. He completed the Clare Hall Study Programme at the University of Cambridge, Cambridge, UK, in 2019. In 2013, he was a Postdoctoral Fellow in The Hong Kong Polytechnic University, Hong Kong, China. He is currently an Associate Professor with the State Key Laboratory of Analog and Mixed-Signal VLSI and the Institute of Microelectronics, UM, and also with the Department of Electrical and Computer Engineering, Faculty of Science and Technology, UM. He has coauthored or co-edited four books and around 200 technical journals and conference papers. He holds six US and five Chinese patents. His research interests include power converters, power quality compensators, photovoltaic energy generation system, power management integrated circuits, and wireless power transfer.

Dr. Lam was the recipient or co-recipient of the IEEE PES Chapter Outstanding Engineer Award in 2016, the Gold Medal of the 48th Geneva International Exhibition of Inventions in 2023, the Macao Science and Technology Invention Award (Second Class and Third Class), and the Macao Postgraduate (Ph.D. Level) Scientific and Technological R&D Award in 2018, 2014, and 2012, respectively. He is currently the Chair of the Young Professionals in Circuits and Systems (YPCAS) of IEEE CASS, Founding Chair of the IEEE Macau IES Chapter, and the Secretary of the IEEE IES Technical Committee on Power Electronics. He currently serves as an Associate Editor for IEEE Transactions on Industrial Electronics, IEEE Transactions on Power Electronics, IEEE Journal of Emerging and Selected Topics in Power Electronics, IEEE Open Journal of the Industrial Electronics Society, and IEEE Access. He served as a Guest Editor for the IEEE Transactions on Circuits and Systems II: Express Briefs (2020–2022) and the IET Power Electronics (2020–2021). He was awarded 2021 IEEE Access Outstanding Associate Editor and the 2022 IEEE Transactions on Industrial Electronics Distinguished Reviewer. He was also the Tutorial Speaker for IECON 2024, IPEMC 2024-ECCE Asia, PEDG 2023, ISCAS 2021, IECON 2020, and APPEEC 2019. Since 2020, he has been listed as the World's Top 2% Scientists by Stanford University for four consecutive years. e-mail: cslam@um.edu.mo; c.s.lam@ieee.org

Cheng Gong is currently a Research Assistant Professor in the State Key Laboratory of Analog and Mixed VLSI, Institute of Microelectronics, University of Macau (UM), Macau, China. He received Ph.D. degree in electrical and computer engineering from UM in 2022 and M.Sc. degree in electrical engineering from the Guangxi University, Nanning, China, in 2017. From January 2022 to July 2023, he worked as a Postdoc Fellow in the State Key Laboratory of Analog and Mixed VLSI, Institute of Microelectronics, UM. His research interest includes power electronics converters, reinforcement learning in power electronics, power quality compensators, and power systems optimization.

Dr. Gong was the recipient or co-recipient of the Gold Medal of the 48th Geneva International Exhibition of Inventions in 2023, UM-Akrostar Academic Award in 2022, and IEEE Macau Section IES Chapter 2020 Postgraduate Paper Contest Championship. He is currently the Chapter Officer of the IEEE Macau IES Chapter and Member of the IEEE IES Technical Committee on Power Electronics. He was also the Tutorial Speaker for IECON 2024 and PEDG 2023. e-mail: cgong@um.edu.mo

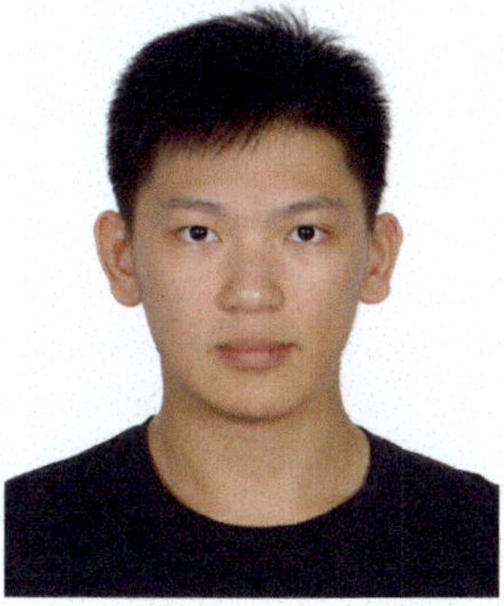

Wai-Kit Sou received the B.Sc. degree and Ph.D. degree in electrical and computer engineering from the University of Macau (UM), Macau, China, in 2019 and 2023, respectively.

Currently, he is a postdoctoral fellow at the Institute of Microelectronics of UM. His current research interests include power electronics, energy saving, and power quality compensation.

Dr. Sou was the recipient or co-recipient of the Macao Scientific and Technological R&D Award 2022 (Ph.D. Level), the Gold Medal of the 48th Geneva International Exhibition of Inventions in 2023. He has been a member in the board of supervisors of PES/PELS Joint Chapter in IEEE Macau Section since 2018. From 2020, he is the secretary of IES Chapter under IEEE Macau Section and member of Power Quality Subcommittee of the IEEE IES PETC. He was also the Tutorial Speaker for IECON 2024 and PEDG 2023. e-mail: wksou@um.edu.mo

Abbreviations

AHCC	Adaptive Hysteresis Current Control
APF	Active Power Filter
BPF	Band-Pass Filter
DBCC	Deadbeat Current Control
FS-HMPC	Finite-Set Hybrid Model Predictive Control
FS-MPC	Finite-Set Model Predictive Control
HAPF	Hybrid Active Power Filter
HCC	Hysteresis Current Control
HOC	H_∞ Optimal Control
HPF	High-Pass Filter
HSS	Harmonic State-Space
IGBT	Insulated Gate Bipolar Transistor
LHP	Left-Hand Plane
LMI	Linear Matrix Inequality
LPF	Low-Pass Filter
LTI	Linear Time Invariant
MPC	Model Predictive Control
MQPRC	Multi-Quasi-Proportional-Resonant Control
PCC	Proportional Current Control
PIC	Proportional Integral Control
PMRC	Proportional Multi-Resonant Control
PPF	Passive Power Filter
PQ	Power Quality
PWM	Pulse Width Modulation
QPRC	Quasi-Proportional-Resonant Control
RHP	Right-Hand Plane
SMC	Sliding Mode Control
SOSMC	Second-Order Sliding Mode Control
SOSMO	Second-Order Sliding Mode Observer
TC*LC*	Thyristor-Controlled *LC*
TDD	Total Demand Distortion

THD	Total Harmonic Distortion
VPIC	Vector Proportional Integral Control
VSI	Voltage Source Inverter

Chapter 1
Introduction

1.1 Power Quality Issues

Nowadays, power electronic devices are indispensable for human life activities and development. Among them, non-linear loads and motor loads, such as converters, bulk rectifier, air conditioning systems and elevators, cause serious pollution on power quality of the power grid. Besides, the proliferation and increased usage of grid-tied renewable energy systems and electric transportation systems also has a crucial impact on power quality in the distribution network [1–7]. According to the survey conducted by the Leonardo Power Quality Initiative and Electric Power Research Institute, the economic loss caused by the poor power quality is estimated to cost annually 15 to 24 billion US dollars in United Sates and more than 150 billion Euros in European Union countries, respectively [8, 9].

1.1.1 Understanding Power Quality Issues

Power quality issues refer to deviations from the ideal characteristics of electrical power, which can adversely affect the performance and reliability of electrical and electronic equipment. These issues encompass a range of phenomena, including voltage sags and swells, harmonics, unbalance current/neutral current, and reactive power [10–12]. Understanding these issues is essential for diagnosing and mitigating problems to ensure the efficient and reliable operation of electrical systems.

Voltage Sags and Swells: Voltage sags are short-duration reductions in RMS voltage, typically caused by sudden increases in load or faults in the power system. Voltage swells, on the other hand, are short duration increases in RMS voltage, often resulting from sudden reductions in load or switching off large loads.

Harmonics: Harmonics are voltage or current waveforms that deviate from the fundamental frequency (typically 50 or 60 Hz). They are generated by non-linear

C.-S. Lam et al., *Advanced Controllers Design for Hybrid Active Power Filters*,
https://doi.org/10.1007/978-981-97-6847-9_1

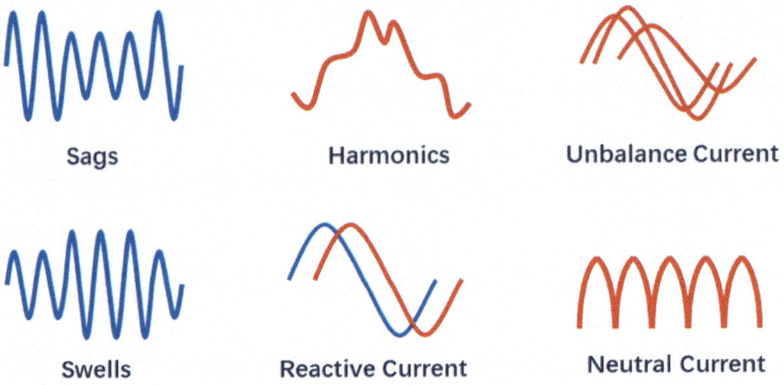

Fig. 1.1 Typical waveforms of the power quality issues

loads such as rectifiers, variable frequency drives, and electronic ballasts, which draw current in a non-sinusoidal manner.

Unbalanced Current/Neutral Current: Unbalanced current occurs when the currents in a three-phase system are not equal in magnitude or are not 120 degrees out of phase, leading to increased losses, overheating, and potential damage to electrical components. This imbalance results in significant neutral current flow, causing additional issues such as increased losses and potential overheating of the neutral conductor, and can also lead to voltage imbalances affecting sensitive equipment.

Reactive Current: Reactive current is the component of the total current that does not contribute to real power consumption, arising from inductive or capacitive elements in the system like motors and capacitors. It leads to power quality issues such as reduced power factor, increased losses, and voltage regulation problems, causing inefficiencies and higher operational costs.

Figure 1.1 illustrates the typical waveforms associated with various power quality issues, providing a visual representation of how these disturbances manifest in electrical systems, which includes examples of voltage sags, swells, harmonic distortions, unbalanced currents, and impulse, each showcasing the unique waveform anomalies that can disrupt the stability and efficiency of power delivery.

1.1.2 Causes of Power Quality Issues

Power quality issues can arise from a variety of sources, both internal and external to the electrical system. Understanding these causes is essential for diagnosing and mitigating power quality problems [10–12].

Voltage Sags and Swells: Often caused by sudden changes in load, such as the startup of large motors, sudden reductions in load, or by faults in the power system like short circuits or a single line-to-ground fault in a grounded system.

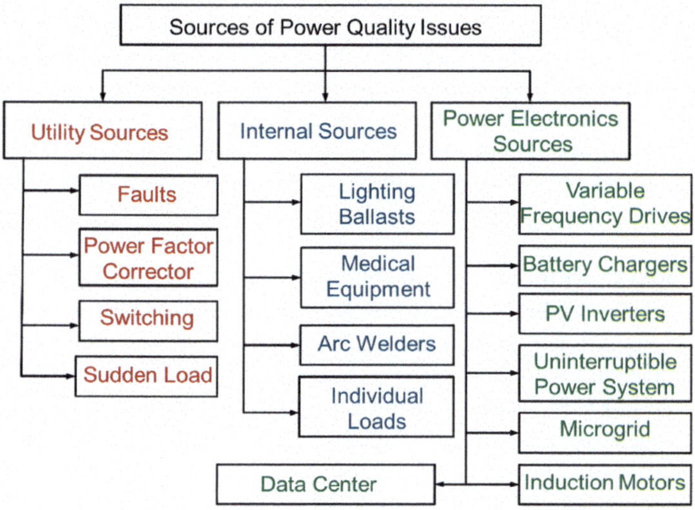

Fig. 1.2 Sources of the power quality issues

Harmonics: Generated by non-linear loads such as rectifiers, variable frequency drives (VFDs), fluorescent lighting, and electronic ballasts. These devices draw current in a non-sinusoidal manner, creating harmonic currents that distort the voltage waveform.

Unbalanced Current/Neutral Current: Unbalanced current is primarily caused by the uneven distribution of single-phase loads across three phases, faults such as open or short circuits, and issues with the supply transformer like unequal winding impedances. Additional contributors include harmonic distortion from non-linear loads and phase angle discrepancies due to improper system design or faults.

Reactive Current: Reactive current arises mainly from inductive and capacitive loads, with inductive loads causing the current to lag and capacitive loads causing it to lead. Other causes include improperly sized or malfunctioning power factor correction equipment, non-linear loads introducing harmonic currents, and system imbalances affecting the phase relationship between voltage and current.

Figure 1.2 provides a comprehensive summary of the sources contributing to various power quality issues, offering a detailed overview of the origins of these disturbances within electrical systems.

1.1.3 Impacts of Power Quality Issues

The hazards associated with poor power quality are numerous and can have significant economic, operational, and safety implications. Some of the key impacts include:

Equipment Damage: Poor power quality can cause physical damage to electrical and electronic equipment. For example, voltage impulse can cause insulation breakdown, while harmonics can lead to overheating of transformers and motors. This damage can result in costly repairs or replacements and can disrupt operations.

Reduced Equipment Lifespan: Continuous exposure to poor power quality conditions can reduce the lifespan of electrical equipment. For instance, voltage sags and swells can cause wear and tear on electronic components, leading to premature failure.

Operational Downtime: Power quality issues can cause equipment malfunctions and unplanned shutdowns, leading to operational downtime. This can be particularly detrimental in industrial and commercial settings, where downtime can result in significant financial losses.

Data Loss and Corruption: Sensitive electronic equipment, such as computers and servers, can be adversely affected by power quality issues. Voltage sags, impulse, and harmonics can cause data loss, corruption, and system crashes, leading to productivity losses and potential data breaches.

Increased Energy Costs: Poor power quality can lead to increased energy consumption and higher operational costs. For example, unbalance and harmonics can cause motors and transformers to operate less efficiently, resulting in higher energy usage.

Safety Risks: Power quality issues can pose safety risks to personnel and equipment. For example, impulse can cause insulation breakdown and arcing, leading to fire hazards. Unbalance can cause motors to overheat, increasing the risk of electrical fires.

1.1.4 Importance of Current Quality Compensation

Among the various power quality issues, the importance of addressing reactive power, harmonics, and unbalanced currents cannot be overstated [12]. Reactive power, which arises from the phase difference between voltage and current, can lead to inefficient power delivery and increased losses within the electrical system. Harmonic currents, generated by non-linear loads such as variable frequency drives and electronic ballasts, introduce waveform distortions that result in excessive heating and reduced efficiency of transformers and motors. Unbalanced currents, often stemming from uneven load distribution or faults, create significant voltage imbalances that further stress the electrical infrastructure.

These issues collectively impact the reliability and efficiency of power systems, leading to increased operational costs, accelerated equipment wear and tear, and potential system failures. Addressing reactive power, harmonics, and unbalanced currents is essential for maintaining the integrity of the power system, ensuring the optimal performance of equipment, and minimizing operational disruptions and maintenance costs. By focusing on these critical aspects of current quality, utilities and consumers can enhance the overall stability and resilience of their electrical systems, thereby promoting a more reliable and efficient power delivery network.

1.1.5 Power Quality Standards

To address power quality issues and ensure the reliable operation of electrical systems, several organizations have established standards and guidelines. These standards provide benchmarks for acceptable power quality levels and offer recommendations for mitigating power quality problems. Some of the prominent organizations that have developed power quality standards [13–18], include the Institute of Electrical and Electronics Engineers (IEEE), the Institution of Engineering and Technology (IET), and the Chinese National Standards (GB). As this book mainly focuses on current quality issues, some of the commonly used standards are briefly introduced in the following:

IEEE Std 519–2014 [14]: The standard addresses the need for controlling harmonic distortion at the point of common coupling (PCC) between the utility and the customer. It provides recommended practices for the design, operation, and maintenance of electrical systems to limit harmonic distortion. The primary focus is on establishing acceptable limits for both harmonic current and voltage distortion to ensure the reliable and efficient operation of electrical equipment. One of the key features of IEEE Std 519–2014 is its detailed guidelines on the acceptable levels of harmonic currents. The standard categorizes harmonic current limits based on the electrical system's voltage level and the short-circuit ratio (SCR) at the PCC. These limits are designed to prevent excessive harmonic distortion that could lead to equipment overheating, increased losses, and potential system failures. The standard specifies maximum allowable levels for individual harmonic currents. For example, for systems with voltages up to 69 kV, the limit for the 5th harmonic is set at 4% of the fundamental current. IEEE Std 519–2014 also provides limits for the total demand distortion (TDD), which is the sum of all harmonic currents expressed as a percentage of the maximum demand load current. For systems with SCR greater than 1000, the TDD limit is set at 5%.

IET Wiring Regulations [16]: The IET Wiring Regulations, known formally as BS 7671, addresses power quality issues, emphasizing the importance of maintaining voltage stability, minimizing harmonic distortion, and protecting against transients. These considerations are crucial for ensuring the reliable operation of sensitive electronic equipment and preventing disruptions in electrical systems. BS 7671 sets forth specific limits for harmonic currents to ensure that the electrical system operates within safe and efficient parameters. These limits are designed to prevent excessive harmonic distortion that could adversely affect both utility infrastructure and consumer equipment. BS 7671 recommends that the Total Harmonic Distortion (THD) of the current should not exceed certain levels to ensure the reliable operation of electrical equipment. For example, the THD of the current should generally be kept below 5% for most installations. The regulations also specify limits for individual harmonic currents. For instance, the 5th harmonic current should be limited to a maximum of 4% of the fundamental current. This helps in controlling the specific harmonic components that are most likely to cause problems in the electrical system.

GB/T 14,549–1993 [17]: Provides guidelines for the measurement and assessment of harmonic distortion in electrical power systems. Defines acceptable limits for harmonic current and voltage distortion and provides recommendations for mitigating harmonic issues. GB/T 14,549–1993 sets forth specific limits for harmonic currents to ensure that the electrical system operates within safe and efficient parameters. These limits are designed to prevent excessive harmonic distortion that could adversely affect both utility infrastructure and consumer equipment. The standard recommends that the Total Harmonic Distortion (THD) of the current should generally be kept below 5% for most installations. This limit helps ensure that the cumulative effect of all harmonic currents does not compromise the performance of electrical systems. GB/T 14,549–1993 specifies limits for individual harmonic currents based on the voltage level of the system. For example, for systems with voltages up to 1 kV, the limit for the 5th harmonic current is set at 4% of the fundamental current. These individual limits help control specific harmonic components that are most likely to cause problems in the electrical system.

1.2 Development of Current Quality Compensators

As early as 1900, capacitor banks have been used to compensate for the current quality problems. However, the CB cannot handle the current harmonics problem. To solve the current harmonics pollution, passive power filter (PPF) was first proposed and installed in the mid-1940s. It was then widely applied to compensate for the reactive power and suppress current harmonics distortion. Unfortunately, both CB and PPF suffer many disadvantages, such as static compensation capability, low dynamic performance, and resonance problems [19]. Since the concept of an "active ac power filter" was first developed by Gyugyi in 1976 [20], research studies on active power filter (APF) for current quality compensation are getting more and more attention due to its fast and dynamic compensation characteristics without resonance problems as in the CB and PPF [21].

Although APFs can overcome the drawbacks of CB and PPF, the initial and operational costs of the APF are relatively high due to its high dc-link operation voltage requirement, this results in slowing down their large-scale application in distribution networks. Hybrid active power filters (HAPFs), which have different properties on power quality compensation, were then introduced in the past two decades to reduce the high costs of the APF and overcome the drawbacks of the PPF [22]. The *LC*-coupling hybrid active power filter (*LC*-HAPF), thyristor-controlled *LC*-coupling hybrid active power filter (TC*LC*-HAPF), and *LCLC*-coupling hybrid active power filter (*LCLC*-HAPF) were proposed in succession for power quality conditioning in the recent two decades [23–25]. In spite of the low DC-link voltage requirement of the *LC*- and *LCLC*-HAPF, their compensation range is limited by the filter structure, resulting in narrow compensation range and not compatible with capacitive loads. Moreover, the *LCLC*-HAPF needs an auxiliary active damping control to dampen the resonance of the filter. The TC*LC*-HAPF has an adaptive filter structure (TC*LC* part), which can widen the reactive power compensation range while keeping a low DC-link voltage requirement. Moreover, it can perform harmonic current mitigation, dynamic reactive and unbalanced power compensation under both inductive and capacitive load [26]. These features make the TC*LC*-HAPF the most cost-effective solution in the medium-voltage distribution power system. The comparison among different active power quality compensators for 11 kV system application has been tabulated in Table 1.1. As Table 1.1 summarized, although the operating range of *LC*-HAPF is not wide, the absence of active devices such as thyristors is still its attractive point compared to TC*LC*-HAPF. The basic topology of APF, *LC*-HAPF, *LCLC*-HAPF and TC*LC*-HAPF are shown in Fig. 1.3. Considering the control complexity, operating range and cost-effectiveness, this book will focus on the current control strategy study of *LC*-HAPF and TC*LC*-HAPF, in which their detailed circuit configuration for this book is given in Figs. 1.4 and 1.5.

The operation principle of HAPFs is depicted in Fig. 1.6. The non-linear loads bring distorted current with harmonic and reactive power in the distribution network. After installing the HAPFs, which are composed of the DC-AC inverter and different coupling filters, it will generate a real-time compensating current to compensate for harmonic current and reactive power. Hence, the source- or system-side current will be compensated to become sinusoidal and in phase with the grid voltage (unity power factor) eventually.

Table 1.1 Comparison among different active power quality compensators for 11 kV system application

	Year	Multi-Level	Control complexity	Operating range	DC-link voltage	Power loss	Cost
APF [20]	1976	**Yes** (3 stages)	**High**	Wide	**High**	**High**	**High**
LC-HAPF [23]	2003	No (1 stage)	Low	**Narrow**	Low	Low	Low
LCLC-HAPF [25]	2020	No (1 stage)	Medium	**Narrow**	Low	Low	Low
TC*LC*-HAPF [24]	2014	No (1 stage)	Low-Medium	Wide	Low	Low-Medium	Medium

Note The bold area means the undesirable characteristic

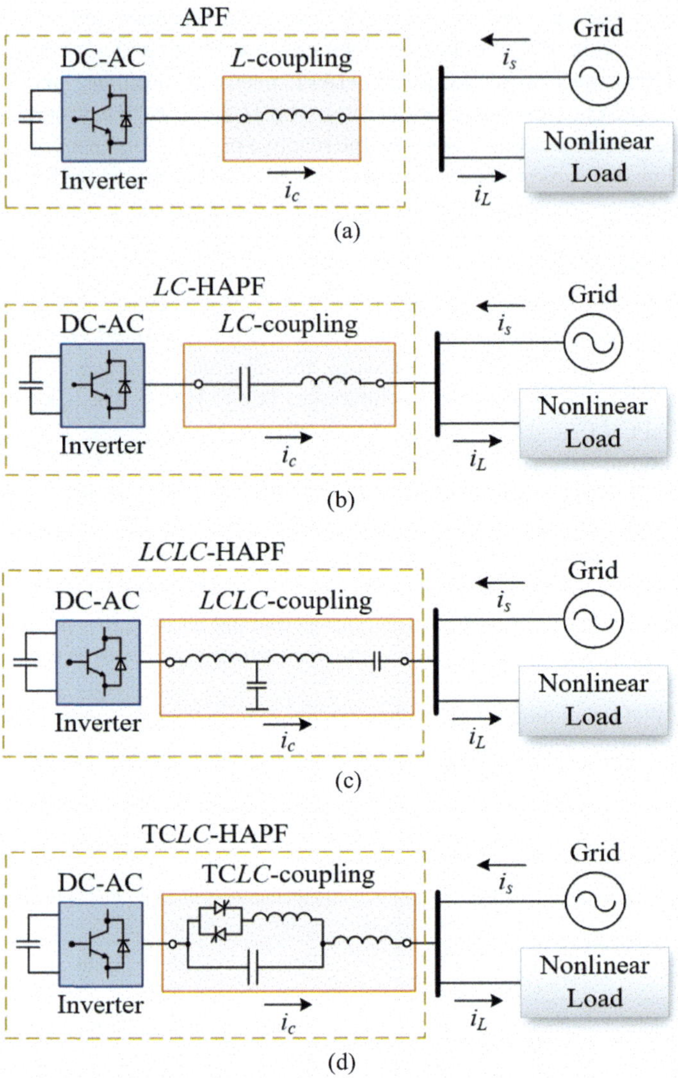

Fig. 1.3 Topology of APF and HAPFs. **a** APF; **b** *LC*-HAPF; **c** *LCLC*-HAPF; **d** TC*LC*-HAPF

1.2.1 Circuit Configuration of LC-*HAPF*

The circuit configuration of a three-phase four-wire center-split *LC*-HAPF is shown in Fig. 1.4, where the subscript '*x*' denotes phase *a*, *b*, *c*, and *n*. v_x is the load voltage and v_{invx} is the inverter voltage. i_{sx}, i_{cx}, and i_{Lx} are the source, compensating and load currents for each phase, respectively. C_{dc} is the DC capacitor which provides the DC-link voltage V_{dc}. T_{1x} and T_{2x} stand for the trigger signal of the insulated-gate

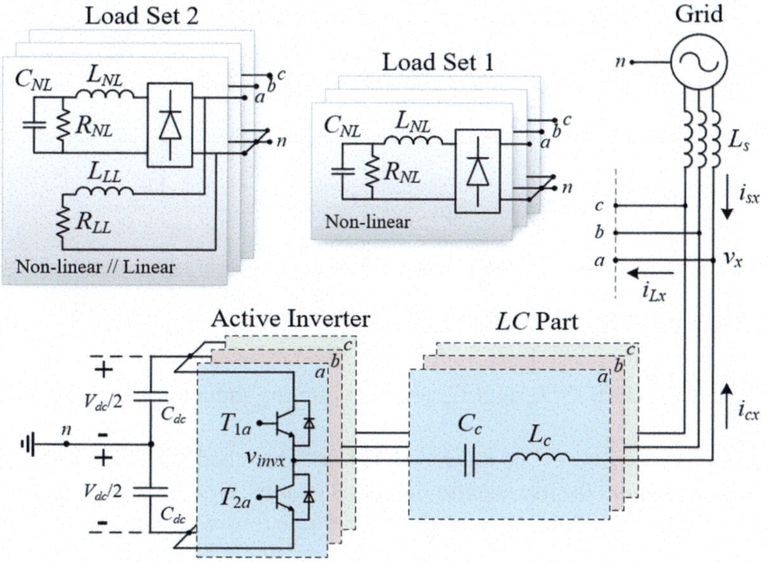

Fig. 1.4 Circuit configuration of *LC*-HAPF

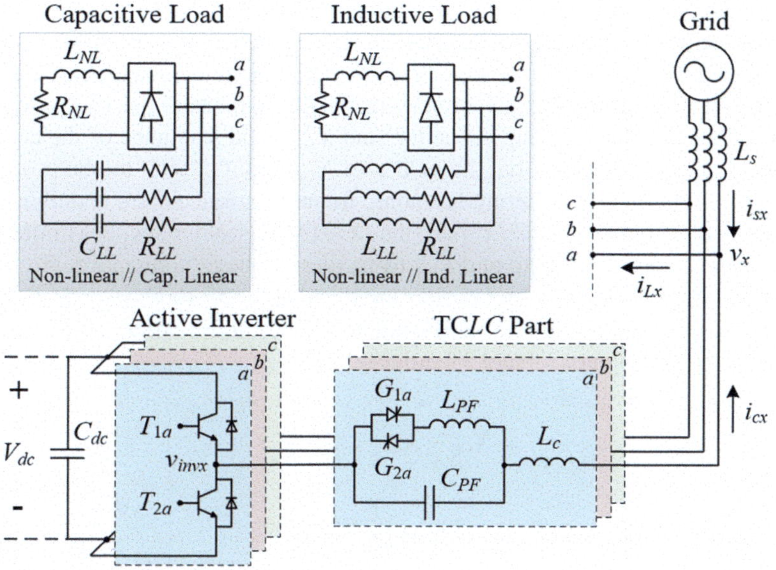

Fig. 1.5 Circuit configuration of TC*LC*-HAPF

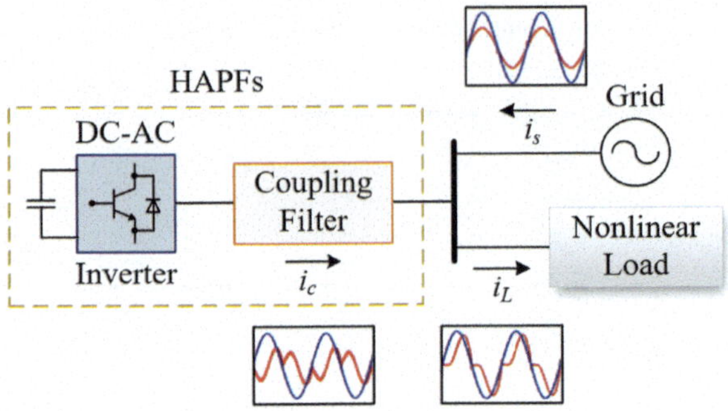

Fig. 1.6 Operation principle of HAPFs

bipolar transistor (IGBT). L_c and C_c are the coupling inductor and capacitor of the LC-HAPF.

The coupling L_c and C_c of the LC-HAPF are designed based on the harmonic current of loading and the average fundamental reactive power consumption. Generally, the dominant harmonic orders in the three-phase four-wire system are odd orders ($n = 3, 5, 7\ldots$) and each LC branch can filter one single harmonic order. More LC branches can be added to the LC-HAPF to filter more harmonic current orders. Otherwise, the minimum dc-link voltage should be increased as discussed in [27–29]. The design criteria of L_c and C_c are described in the following.

The relationship between the load fundamental reactive power and reactance of L_c and C_c can be expressed in (1.1), where ω_1 is the fundamental angular frequency, V_x represents the root-mean-square load voltage, Q_{Lx} is the load reactive power, X_{Cc} is the reactance of the coupling capacitor, and X_{Lc} is the reactance of the coupling inductor. After that, it can solve the values of X_{Cc} and X_{Lc} with the help of (1.2). Note that n is the order of filtering harmonic frequency. Since there is only a single tuned passive LC branch in the LC part, the LC-HAPF is suitable under limited load reactive power variation range and low-to-medium voltage application. To reduce the cost and size of the coupling LC, they are tuned at the 5th order ($n = 5$) harmonic frequency in this book because the dominant harmonic current order of the full bridge diode rectifier loading are 3rd and 5th orders.

$$\frac{V_x^2}{|Q_{Lxf}|} = X_{Cc} - X_{Lc} = \frac{1}{\omega_1 C_c} - \omega_1 L_{cx} \tag{1.1}$$

$$X_{Lc} = \frac{1}{n^2} X_{Cc} \tag{1.2}$$

1.2.2 Circuit Configuration of TCLC-HAPF

The system configuration shown in Fig. 1.5 is a three-phase three wire TC*LC*-HAPF, which is composed by TC*LC* part and active inverter part. The subscript '*x*' represents phase *a*, *b*, *c*. v_x, v_{invx} represent the load and inverter voltage. i_{sx}, i_{Lx}, i_{cx} represent the source, load and compensating currents, respectively. C_{dc} is the DC capacitor which provides the DC-link voltage V_{dc}. T_{1x} and T_{2x} stand for the trigger signal of the IGBT and G_{1x} and G_{2x} are the trigger signal of the thyristor.

The TC*LC* part can provide a wide and continuous reactive power compensation range under inductive, capacitive or unbalanced loads by controlling two thyristors of the TC*LC* part in each phase, which means it is more capable of working under the wide load reactive power variation range and medium voltage scenario. It is composed of a ripple filtering inductor L_c and a thyristor-controlled reactor L_{PF} connected in parallel with a coupling capacitor C_{PF}. The purpose of the L_c is to filter out the current ripple caused by the IGBTs and it is designed by [30]:

$$L_c \geq \frac{V_{dc}}{8 \cdot f_{sw} \cdot \Delta i_{Lc\max}} \tag{1.3}$$

where f_{sw} is the inverter switching frequency, and $\Delta i_{Lc\max}$ is the maximum allowed compensating current ripple value.

The L_{PF} and C_{PF} determine the range of reactive power compensation of the TC*LC* part. Their value need to fulfill the following two inequalities [30]:

$$\frac{1}{\omega_1^2 L_c} > C_{PF} \geq \frac{1}{\omega_1^2 L_c - \omega_1 X_{\mathrm{TCLC}}} \tag{1.4}$$

$$L_{PF} \leq \frac{X_{\mathrm{TCLC}} - \omega_1 L_c}{\omega_1^2 X_{\mathrm{TCLC}} C_{PF} - \omega_1^3 L_c C_{PF} + \omega_1} \tag{1.5}$$

where the ω_1 is the fundamental angular frequency and X_{TCLC} is the fundamental reactance of the TC*LC* part. The relationship between the X_{TCLCx} and the load reactive power Q_{Lx} can be expressed as (1.6), where $\overline{V}_x$ is the rms value of the load voltage.

$$\begin{bmatrix} X_{TCLCa} \\ X_{TCLCb} \\ X_{TCLCc} \end{bmatrix} = \begin{bmatrix} \frac{3\overline{V}_x^2 (Q_{Lc}-Q_{Lb}-Q_{La})^{-1}(Q_{Lb}-Q_{La}-Q_{Lc})^{-1}}{(Q_{Lc}-Q_{Lb}-Q_{La})^{-1}+(Q_{La}-Q_{Lb}-Q_{Lc})^{-1}+(Q_{Lb}-Q_{La}-Q_{Lc})^{-1}} \\ \frac{3\overline{V}_x^2 (Q_{Lc}-Q_{Lb}-Q_{La})^{-1}(Q_{La}-Q_{Lb}-Q_{Lc})^{-1}}{(Q_{Lc}-Q_{Lb}-Q_{La})^{-1}+(Q_{La}-Q_{Lb}-Q_{Lc})^{-1}+(Q_{Lb}-Q_{La}-Q_{Lc})^{-1}} \\ \frac{3\overline{V}_x^2 (Q_{La}-Q_{Lb}-Q_{Lc})^{-1}(Q_{Lb}-Q_{La}-Q_{Lc})^{-1}}{(Q_{Lc}-Q_{Lb}-Q_{La})^{-1}+(Q_{La}-Q_{Lb}-Q_{Lc})^{-1}+(Q_{Lb}-Q_{La}-Q_{Lc})^{-1}} \end{bmatrix} \tag{1.6}$$

The purpose of the active inverter part is to compensate for the harmonics current and intensify the dynamic compensation capability of the TC*LC*-HAPF. Its structure is a 2-level DC/AC voltage source inverter (VSI) connected with the DC-link capacitor C_{dc}.

1.3 Development of Control Strategies for HAPFs

In addition to the structure, the controller is also a crucial part that determines the performance of HAPFs. The following sections provide a development summary on the control strategies for HAPFs.

1.3.1 Control Strategies for LC*-HAPF*

Conventionally, a proportional current controller (PCC) is applied for the *LC*-HAPF [23], but the steady-state error is significant. Since 2012, hysteresis current control (HCC) and adaptive hysteresis current control (AHCC) have been developed for the *LC*-HAPF [31, 32], respectively, which provides a good dynamic performance. However, the HCC has a varying switching frequency which leads to undesired high-order harmonics in a wide frequency range, and its steady-state error is still significant. In 2016, a quasi-proportional-resonant control (QPRC) is proposed [33], which achieves low steady-state error at the fundamental frequency. However, the steady-state error of the QPRC at harmonic frequencies is significant. The low steady-state error and the selective harmonic compensation can be achieved by the multi-quasi-proportional-resonant control (MQPRC) or proportional multi-resonant control (PMRC), which has been applied to the *LC*-HAPF in [34]. Nevertheless, the transient response speed of the *LC*-HAPF with the MQPRC is slow because of the integration process. Therefore, the former existing *LC*-HAPF current controllers suffer from one or more drawbacks of high steady-state error, slow dynamic response, and varying frequency.

In this book, a deadbeat current control (DBCC) is first proposed for the *LC*-HAPF. DBCC is characterized by fixed switching frequency, good steady-state performance, fast dynamic response, and relatively simple control, thus leading it to become an attractive control strategy for the APF in the industry [35–37]. As a result, it is valuable to develop the DBCC for the *LC*-HAPF.

Model predictive control (MPC) has already been applied in the process and chemical industry successfully for several decades [38]. And it has been applied for the APF [39, 40] and other power converter applications [40–43] so far. Hence, a model predictive control (MPC) and improved MPC for the *LC*-HAPF are proposed in this book, which can obtain good steady-state performance and fast dynamic response and because it directly selects the optimal switching state of the converter and does not require any modulator for the PWM signal generation process [38]. The added signal correction term of improved MPC further enhances the steady-state performance compared to the convention MPC.

The aforementioned two kinds of controller are model-based control methods, they all require a coupling capacitor voltage sensing circuit. In order to save hardware costs, other non-model-based control methods are also proposed in this book. A vector proportional-integral control (VPIC) for the *LC*-HAPF with low steady-state

Table 1.2 Comparison among different current control strategies for *LC*-HAPF

Controller	Year	Design Complexity	Steady-state Accuracy	Transient Response Speed	Switching Frequency
PCC [23]	2003	Low	**Low**	Medium	Fixed
HCC [31]	2012	Low	Medium	Fast	**Varying**
QPRC [33]	2016	Medium	**Low**	Medium	Fixed
MQPRC [34]	2016	**High**	High	**Slow**	Fixed
DBCC	This book	Low	Medium	Fast	Fixed
Improved MPC	This book	Medium	Medium	Fast	**Varying**
VPIC	This book	Medium	High	Medium	Fixed
SOSMC	This book	Medium	High	Fast	Fixed

Notes The bold areas mean undesirable performance

error, fast transient response, fixed switching frequency, and good robustness even under a minimum DC-link voltage is proposed. The coupling term (plant model) of the control system can be cancelled by the VPIC, which improve the performance in terms of transient, stability and robustness compared to the conventional resonant control method [44, 45].

Finally, a second-order sliding mode control (SOSMC), which achieve a low steady-state error, fast transient response, fixed switching frequency, good robustness under system parameters variation, and good grid disturbance rejection capability under non-ideal grid conditions, is proposed for the *LC*-HAPF. SOSMC is the simplest and the most widely used high-order sliding mode control (SMC). It has been developed recently for controlling the active filter [46], and APFs [47] since it solves the variable switching frequency problem of the conventional SMC (HCC is a special case of SMC) [48] and the robustness is very strong. Therefore, developing SOSMC for the *LC*-HAPF is highly valuable.

The detailed design process and validation results of DBCC, improved MPC, VPIC and SOSMC for the *LC*-HAPF can be found in Ch. 2–5. Table 1.2 provides a brief comparison among the current control strategies for the *LC*-HAPF.

1.3.2 Control Strategies for TCLC-HAPF

Proportional-integral control (PIC) [24] was the first controller applied to the TC*LC*-HAPF. However, by applying the linear PIC to the nonlinear TC*LC*-HAPF system, due to its nonlinear coupling TC*LC* part, three additional voltage sensors are required for linearizing the TC*LC*-HAPF system model. Besides, the PIC also suffers from poor system robustness and disturbance rejection capability. Hysteresis current control (HCC) is usually used for controlling the TC*LC*-HAPF [26, 30, 49], because of its simplicity and fast dynamic response. However, due to the digital control delay and the limited sampling frequency in the actual implementation, the hysteresis band of HCC cannot be set as infinite small to guarantee the steady-state performance. And the switching frequency of the HCC is varying that leads to undesired current ripple in a wide frequency range. Therefore, the existing TC*LC*-HAPF control strategies still need to be studied to overcome the nonlinearity of coupling TC*LC* part, achieve low steady-state error and fast transient response.

The multi-quasi-proportional-resonant control (MQPRC) has already been applied in *LC*-HAPF [34]. However, the filter (TC*LC* part) model of the TC*LC*-HAPF is not linear unlike those of the APF (*L* model) [20] and the *LC*-HAPF [23], which means that the design of the MQPRC for the TC*LC*-HAPF under linear control model consideration should be further studied and investigated. In this book, a MQPRC for the TC*LC*-HAPF is proposed based on the separate equivalent model of the TC*LC* part. Through the gain scheduling algorithm for the resonant gain at the fundamental frequency, the MQPRC can improve the steady-state performance under both inductive and capacitive loads conditions.

The harmonic state space (HSS) model of the TC*LC*-HAPF is developed in this book to overcome the nonlinearity of the TC*LC* part. Based on the HSS modeling technique [50, 51], the nonlinear TC*LC*-HAPF is modeled as a linear time-invariant (LTI) system with high accuracy. To obtain the optimal control performance, linear matrix inequality (LMI) based H_∞ optimal controller (HOC) [52] is proposed for the TC*LC*-HAPF.

As mentioned in 1.3.1, many attentions have been drawn by model predictive control (MPC) in the power electronics in the recent years [38, 39]. As one important branch of the MPC, finite-set MPC (FS-MPC) is also valuable to applied for the TC*LC*-HAPF. Hence, a FS-MPC is proposed for the TC*LC*-HAPF in this book. The coupling capacitor voltage observer is utilized in the FS-MPC to reduce the hardware costs. Also, the predictive model simplification is proposed for overcoming the nonlinearity of the TC*LC* part and reducing the required sensing signals.

By combining the SMC and FS-MPC, a finite-set hybrid model predictive control (FS-HMPC) is proposed in this book. This control method utilizes the advantages of SMC and FS-MPC, which improves the robustness of the FS-MPC. Therefore, the FS-HMPC can achieve a superb performance in the aspect of robustness, steady-state error and transient response.

Last but not least, a second order sliding mode control (SOSMC) is also proposed for the TC*LC*-HAPF. As introduced in 1.3.1, the SOSMC can achieve low steady-state error, fast transient response, fixed switching frequency, good system robustness under system parameters variation, and grid disturbance conditions. Moreover, a second-order sliding mode observer (SOSMO) is proposed for the TC*LC*-HAPF in this book to estimate the coupling voltage observer.

The detailed design process and validation results of MQPRC, HOC, FS-MPC, FS-HMPC and SOSMC for the TC*LC*-HAPF can be found in Chaps. 6–10. Table 1.3

Table 1.3 Comparison among different current control strategies for TC*LC*-HAPF

Controller	Year	Design complexity	Steady-state accuracy	Transient response speed	Switching frequency
PIC [24]	2014	Medium	Medium	Medium	Fixed
HCC [49]	2016	Low	**Low**	Fast	**Varying**
MQPRC	This book	**High**	High	**Slow**	Fixed
HOC	This book	**High**	High	Medium	Fixed
FS-MPC	This book	Medium	Medium	Fast	**Varying**
FS-HMPC	This book	Medium	High	Fast	**Varying**
SOSMC	This book	Medium	High	Fast	Fixed

Notes The bold areas mean undesirable performance

provides a brief comparison among the current control strategies for the TC*LC*-HAPF.

1.4 Book Organization

This book is organized into 11 chapters. This chapter presents an overview of the power quality issues, the literature reviews about the structure of HAPFs and their current control strategies. From Chaps. **2** to **5**, the advanced current controllers design of *LC*-HAPF are proposed, while the design of advanced current controllers for TC*LC*-HAPF are presented from Chaps. **6** to **10**. The final chapter will draw conclusions and discuss the future works.

Chapter 2 presents the DBCC design for the *LC*-HAPF, which can concurrently achieve low steady-state error and fast transient response. The control concept and mathematical modeling will be given. After that, the parameter design process based on the stability analysis will be presented. The overall control block diagram and representative results will be given.

Chapter 3 presents the design of improved MPC with signal correction technique for *LC*-HAPF, which can reduce the steady-state error due to discretization and prediction compared to the conventional MPC, hence improve the steady-state performance. The control objective and model deduction will be given. And the parameter design of the signal correction term will be discussed. The overall control block diagram and results compared to conventional MPC and other control methods will also be given.

Chapter 4 presents the VPIC design that provides ideally zero steady-state error and optimal transient response with a detailed system performance and robustness analysis. The digital implementation method of the VPIC and an overall control block diagram for the *LC*-HAPF with the proposed VPIC are presented. Moreover, a comparative analysis between the VPIC and the other representative current controllers under the minimum DC-link operating voltage is discussed.

Chapter 5 presents the SOSMC design for the *LC*-HAPF. The control objective and sliding surface design are proposed. Based on the control objective, the SOSMC parameter design methodology under system stability, system parameter variation, distorted grid voltage, and grid frequency variation consideration is provided. With the help of super-twisting controller, the SOSMC can achieve fixed switching frequency characteristic.

Chapter 6 presents the MQPRC with gain scheduling algorithm for the TC*LC*-HAPF. It can improve the steady-state performance under both inductive and capacitive loads conditions. The modeling of the TC*LC* part, 5 parameters design procedure, overall control block diagram and results will be presented in this chapter.

Chapter 7 presents the HSS model for the TC*LC*-HAPF to address its nonlinearity issue. The HSS based TC*LC*-HAPF model is linear time-invariant (LTI), which enables the optimal control techniques. Then, the linear matrix inequality (LMI) based HOC design is proposed, which finds the optimal controller parameters via convex optimization algorithm. The HOC for the TC*LC*-HAPF achieves high steady-state accuracy, fast transient response, and fixed switching frequency simultaneously.

Chapter 8 presents the FS-MPC with coupling capacitor voltage observer for the TC*LC*-HAPF. The coupling capacitor voltage observer is utilized to reduce the hardware costs when implementing the FS-MPC with proposed simplified predictive model. A comprehensive design process, overall control block diagram and results will be given in this chapter.

Chapter 9 presents the FS-HMPC for the TC*LC*-HAPF. The combined SMC and FS-MPC predictive sliding surface construction will be given. The sliding coefficient tuning method, stability and robustness analysis and overall control block diagram will be presented in detail. Also, the simulation and experimental results will be given.

Chapter 10 presents the SOSMO based SOSMC for the TC*LC*-HAPF. The SOSMO for the TC*LC*-HAPF with its parameter design is first provided. Based on the SOSMO model, the SOSMC for the TC*LC*-HAPF is proposed and designed accordingly. The proposed SOSMO based SOSMC provides low steady-state error, fast dynamic response, fixed switching frequency, and good system robustness under parameter variation and non-ideal grid conditions.

Chapter 11 finally draws the relevant concluding remarks and proposes prospective future research works.

References

1. R.C. Dugan, M.F. McGranaghan, S. Santoso, H.W. Beaty, *Electrical Power Systems Quality* (McGraw-Hill, New York, 1996)
2. P. Salmerón, S.P. Litrán, A Control Strategy for Hybrid Power Filter to Compensate Four-Wires Three-Phase Systems. IEEE Trans. Power Electron. **25**(7), 1923–1931 (2010)
3. P. Salmeron, S.P. Litran, Improvement of the Electric Power Quality Using Series Active and Shunt Passive Filters. IEEE Trans. Power Del. **25**(2), 1058–1067 (2010)
4. J. Yu, Y. Xu, Y. Li, Q. Liu, An inductive hybrid UPQC for power quality management in premium-power-supply-required applications. IEEE Access **8**, 113342–113354 (2020)
5. S. Martinenas, K. Knezović, M. Marinelli, Management of power quality issues in low voltage networks using electric vehicles: Experimental validation. IEEE Trans. Power Del. **32**(2), 971–979 (2017)
6. V.L. Srinivas, B. Singh, S. Mishra, Enhanced power quality PV inverter with leakage current suppression for three-phase SECS. IEEE Trans. Ind. Electron. **69**(6), 5756–5767 (2022)
7. S. Devassy, B. Singh, Performance analysis of solar PV array and battery integrated unified power quality conditioner for microgrid systems. IEEE Trans. Ind. Electron. **68**(5), 4027–4035 (2021)
8. R. Targosz, J. Manson, Pan-European power quality survey. 2007 9th International Conference on Electrical Power Quality and Utilisation, pp. 1–6 (2007)
9. Primen. The cost of power disturbances to industrial and digital economy companies. Report no. 3002000476 (Available through EPRI). Madison, Wisconsin. Primen. (2001)
10. M.H. Bollen, Understanding power quality problems. IEEE, (1999)
11. A. Ghosh, G. Ledwich, *Power quality enhancement using custom power devices* (Springer, New York, NY, 2012)
12. H. Akagi, E.H. Watanabe, M. Aredes, Instantaneous power theory and applications to power conditioning, 2nd ed. Hoboken, NJ: Wiley-Blackwell, (2017)
13. IEEE recommended practice for monitoring electric power quality. IEEE Std 1159–2019 (Revision of IEEE Std 1159–2009), **13**, (2019) pp. 1–98
14. IEEE recommended practice and requirements for harmonic control in electric power systems, IEEE Std 519–2014 (Revision of IEEE Std 519–1992), (2014). pp.1–29
15. IEEE recommended practice for the analysis of fluctuating installations on power systems. IEEE Std 1453–2015 (Revision of IEEE Std 1453–2011). pp. 1–74 (2015)
16. Darrell Locke, Guide to the wiring regulations 17th edition IET wiring regulations (BS 7671: 2008), John Wiley & Sons Ltd, (2008)
17. "Quality of electric energy supply-harmonic in public supply network," Standardization Admin. China, Beijing, China, Tech. Rep. GB/T 14549–93, (1993)
18. "Electricity quality supply voltage deviation," Standardization Admin. China, Beijing, China, Tech. Rep. GB/T 12325–08, (2008)
19. J.C. Das, Passive filters—Potentialities and limitations. IEEE Trans. Ind. Appl. **40**(1), 232–241 (2004)
20. F.Z. Peng, H. Akagi, A. Nabae, A new approach to harmonic compensation in power systems—A combined system of shunt passive and series active filters. IEEE Trans. Ind. Appl. **26**(6), 983–990 (1990)
21. H. Hu, Y. Xing, Design considerations and fully digital implementation of 400-Hz active power filter for aircraft applications. IEEE Trans. Ind. Electron. **61**(8), 3823–3834 (2014)
22. B. Singh, V. Verma, A. Chandra, K. Al-Haddad, Hybrid filters for power quality improvement. IEE Proc. - Gener. Transm. Distri. **152**(3), 365 (2005)
23. S. Srianthumrong, H. Akagi, A medium-voltage transformerless AC/DC power conversion system consisting of a diode rectifier and a shunt hybrid filter. IEEE Trans. Ind. Appl. **39**(3), 874–882 (2003)
24. S. Rahmani, A. Hamadi, K. Al-Haddad, L.A. Dessaint, A combination of shunt hybrid power filter and thyristor-controlled reactor for power quality. IEEE Trans. Ind. Electron. **61**(5), 2152–2164 (2014)

25. Z. Xiang, Y. Pang, L. Wang, C.-K. Wong, C.-S. Lam, M.-C. Wong, Design, control and comparative analysis of an LCLC coupling hybrid active power filter. IET Power Electron. **13**(6), 1207–1217 (2020)
26. L. Wang, C.-S. Lam, M.-C. Wong, Selective compensation of distortion, unbalanced and reactive power of a thyristor-controlled LC-coupling hybrid active power filter (TCLC-HAPF). IEEE Trans. Power Electron. **32**(12), 9065–9077 (2017)
27. C.-S. Lam, X.-X. Cui, W.-H. Choi, M.-C. Wong, Y.-D. Han, Minimum inverter capacity design for three-phase four-wire LC-hybrid active power filters. IET Power Electron. **5**(7), 956–968 (2012)
28. C.-S. Lam, W.-H. Choi, M.-C. Wong, Y.-D. Han, Adaptive DC-link voltage controlled hybrid active power filters for reactive power compensation. IEEE Trans. Power Electron. **27**(4), 1758–1772 (2012)
29. C.-S. Lam, M.-C. Wong, *Design and Control of Hybrid Active Power Filters*, Springer, (2014)
30. L. Wang, C.-S. Lam, M.-C. Wong, Modeling and parameter design of thyristor-controlled LC-coupled hybrid active power filter (TCLC-HAPF) for unbalanced compensation. IEEE Trans. Ind. Electron. **64**(3), 1827–1840 (2017)
31. C.-S. Lam, M.-C. Wong, Y.-D. Han, Hysteresis current control of hybrid active power filters. IET Power Electron. **5**(7), 1175–1187 (2012)
32. L. Wang, C.-S. Lam, M.-C. Wong, An adaptive hysteresis band controller for LC-coupling hybrid active power filter with approximate constant switching frequency. Proc. IEEE PES Asia–Pacific Power Energy Eng. Conf., pp. 1–5 (2014)
33. T. Ye, N. Dai, C.-S. Lam, M.-C. Wong, J.M. Guerrero, Analysis, Design, and Implementation of a Quasi-proportional-resonant controller for a multifunctional capacitive-coupling grid-connected inverter. IEEE Trans. Ind. Appl. **52**(5), 4269–4280 (2016)
34. Z. He, F. Ma, M. Li, Q. Xu, F. Zhou, Y. Yue, Multi-functional hybrid active power converter and its industrial application for electrolytic copper-foil. IET Gener. Transm. Distrib. **10**(16), 3985–3994 (2016)
35. D.G. Holmes, D.A. Martin, Implementation of a direct digital predictive current controller for single and three phase voltage source inverters. In Proc. Conf. Rec. IEEE-IAS Annu. Meeting, pp. 906–913 (1996)
36. M. Odavic, V. Zanchetta, P. Sumner, M.M. Degano, One-sampleperiod-ahead predictive current control for high-performance active shunt power filters. IET Power Electron. **4**(4), 414–423 (2011)
37. W. Jiang, X. Ding, Y. Ni, J. Wang, L. Wang, W. Ma, An improved deadbeat control for a three-phase three-line active power filter with current-tracking error compensation. IEEE Trans. Power Electron. **33**(3), 2061–2072 (2017)
38. Y. Yang, J. Pan, H. Wen, R. Na, H. Wang, Z. Zhang, Z. Ke, L. Xu, An optimized model predictive control for three-phase four-level hybrid-clamped converters. IEEE Trans. Power Electron. **35**(6), 6470–6481 (2020)
39. P. Falkowski, A. Sikorski, Finite control set model predictive control for grid-connected AC–DC converters with LCL filter. IEEE Trans. Ind. Electron. **65**(4), 2844–2852 (2018)
40. X. Chen, W. Wu, N. Gao, H.S.-H. Chung, M. Liserre, F. Blaabjerg, Finite control set model predictive control for LCL-filtered grid-tied inverter with minimum sensors. IEEE Trans. Ind. Electron. **67**(12), 9980–9990 (2020)
41. C. Zheng, T. Dragičević, Z. Zhang, J. Rodriguez, F. Blaabjerg, Model predictive control of LC-filtered voltage source inverters with optimal switching sequence. IEEE Trans. Power Electron. **36**(3), 3422–3436 (2021)
42. M. Easley, M.B. Shadmand, H. Abu-Rub, "Hierarchical model predictive control of grid-connected cascaded multilevel inverter", *IEEE Trans.* Emerg. Sel. Topics Power Electron. **9**(3), 3137–3149 (2021)
43. T. Jin, X. Shen, T. Su, R.C.C. Flesch, Model predictive voltage control based on finite control set with computation time delay compensation for PV systems. IEEE Trans. Energy Convers. **34**(1), 330–338 (2019)

44. H.B.A. Majeed, A. Khan, W. Iqbal, H.A. Khalid, Vector proportional integral controller in series connected custom power device for voltage quality enhancement. In: 2021 International Conference on Emerging Power Technologies (ICEPT), Topi, Pakistan, 2021, pp. 1–5.
45. A.G. Yepes, A. Vidal, J. Malvar, O. Lopez, J. Doval-Gandoy, Tuning method aimed at optimized settling time and overshoot for synchronous proportional-integral current control in electric machines. IEEE Trans. Power Electron. **29**(6), 3041–3054 (2014)
46. J. Liu, S. Vazquez, L. Wu, A. Marquez, H. Gao, L.G. Franquelo, Extended state observer-based sliding-mode control for three-phase power converters. IEEE Trans. Ind. Electron. **64**(1), 22–31 (2017)
47. J. Ramirez, A. Castro, P. Zuniga, A.Y. Alanis, High order sliding mode control for shunt active power filter. Proc. IEEE Int. Autumn Meeting Power, Electron. Comput., pp. 1–6 (2015)
48. V.I. Utkin, *Sliding Modes in Control and Optimization* (Springer, Berlin, Germany, 2013)
49. L. Wang, C.-S. Lam, M.-C. Wong, A hybrid-STATCOM with wide compensation range and low DC-link voltage. IEEE Trans. Ind. Electron. **63**(6), 3333–3343 (2016)
50. J.R.C. Orillaza, A.R. Wood, Harmonic State-Space Model of a Controlled TCR. IEEE Trans. Power Deliv. **28**(1), 197–205 (2013)
51. Kwon, X. Wang, F. Blaabjerg, C.L. Bak, V.-S. Sularea, C. Busca, Harmonic interaction analysis in a grid-connected converter using harmonic State-Space (HSS) Modeling, IEEE Trans. Power Electron., **32**(9), pp. 6823–6835, (2017)
52. S. Boyd, L. El Ghaoui, E. Feron, V. Balakrishnan, Linear matrix inequalities in system and control theory. SIAM, (1994)

Chapter 2
Deadbeat Current Control for *LC*-HAPF

2.1 Introduction

In this chapter, a deadbeat current control (DBCC) for *LC*-HAPF is proposed, which can track with the reference compensation current with low steady-state error and fast dynamic response. Moreover, it can lead *LC*-HAPF to be operating at a fixed switching frequency with low output current ripples, thus reducing the size of the filtering circuit. First of all, the mathematical model of the deadbeat current controller for the *LC*-HAPF is deduced based on its single-phase equivalent circuit. Then, the closed-loop control block diagram and its s-domain model is proposed and built. After that, the stability issue and parameter design of the proposed deadbeat current controller is analyzed and discussed. Finally, the compensating performance of the deadbeat current controller for the *LC*-HAPF is verified by simulation and experimental results compared with other current controllers for the *LC*-HAPF, which shows its effectiveness and superior compensating performances.

The main contribution of this chapter is highlighted in below:

(1) to present a deadbeat current control strategy for *LC*-HAPF to reduce the steady-state error of the compensating current and output current ripples based on the deduced mathematical model and control block diagram.
(2) to study the stability issue and parameter design of the deadbeat current controller for the *LC*-HAPF to ensure its stability.
(3) to verify the proposed deadbeat current controller for the *LC*-HAPF by simulation and experimental results in comparison with the conventional hysteresis current controller, the PI controller and the PMR controller.

C.-S. Lam et al., *Advanced Controllers Design for Hybrid Active Power Filters*,
https://doi.org/10.1007/978-981-97-6847-9_2

2.2 Control Concept

The conventional hysteresis current control (HCC) for the *LC*-HAPF will cause significant steady-state error and also the switching frequency is variable, which implies complexity in design of the filtering circuits and the stable feedback controller for wide range operation. Thus, HCC usually generates large ripple and deteriorates the current harmonics compensation performance of the *LC*-HAPF. To resolve the aforementioned problems, a DBCC will be proposed in this section. Figure 2.1 shows a control block diagram of the proposed DBCC while Fig. 2.2 represents its *s*-domain model, in which the further analysis and discussion of the design of the controller will be presented in the following. Figure 2.3 also shows its working principle waveforms of generating PWM trigger signals. The key concept of this controller is to find out the duty ratio d_x of the switching devices (i.e. inverter voltage v_{invx}) in every switching period which is fixed in every single period ($T_{sw} = 1/f_{sw}$) based on the *LC*-HAPF system parameters, sampling period T_s and sensed instantaneous load voltage v_x, compensation current i_{cx} and coupling capacitor voltage v_{cx} signals. After that, the trigger PWM signals for the switching devices can be generated by comparing the computed d_x with a triangular wave carrier with a fixed switching frequency f_{sw}.

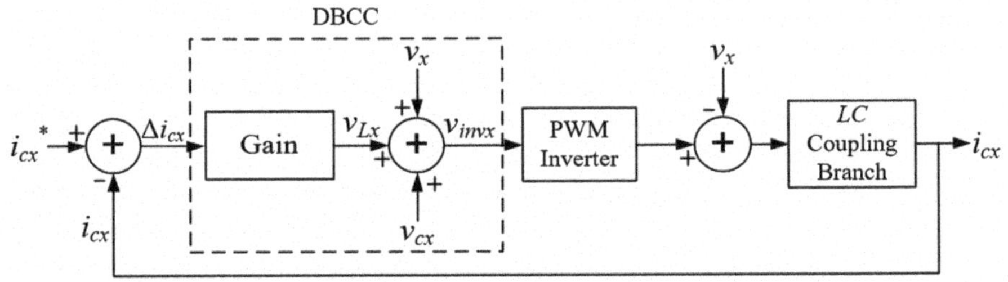

Fig. 2.1 Control block diagram of deadbeat current control

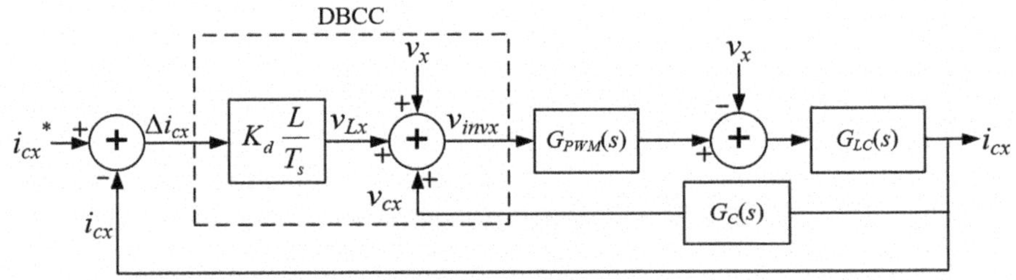

Fig. 2.2 *S*-domain control block diagram model of deadbeat current control

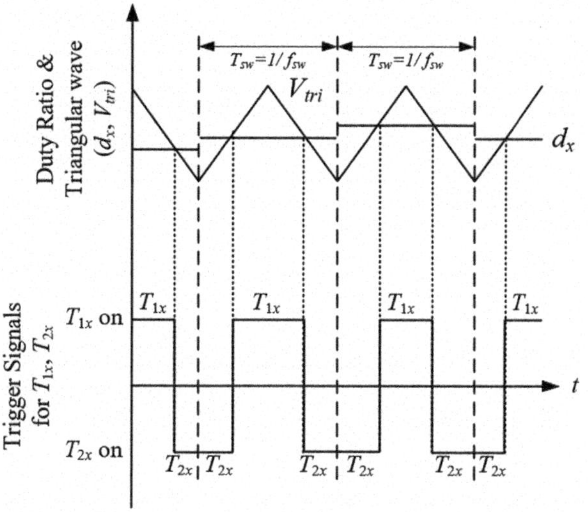

Fig. 2.3 Working principle waveforms of generating PWM trigger signals with DBCC

2.3 Design of Deadbeat Current Control of *LC*-HAPF

2.3.1 *Mathematical Modeling*

The reference compensating current i_{cx}^* can be obtained by using the instantaneous *p-q* theory [1]. Based on Fig. 1.4, the *LC*-HAPF single-phase equivalent circuit can be depicted as Fig. 2.4. From that, the characteristic equation of the *LC*-HAPF system can be expressed as:

$$v_{Lx}(t) = L_{cx}\frac{di_{cx}(t)}{dt} = v_{invx}(t) - v_x(t) - v_{Cx}(t) \tag{2.1}$$

Reforming (2.1), the averaged model of the *LC*-HAPF in the stationary coordinate system can be represented as:

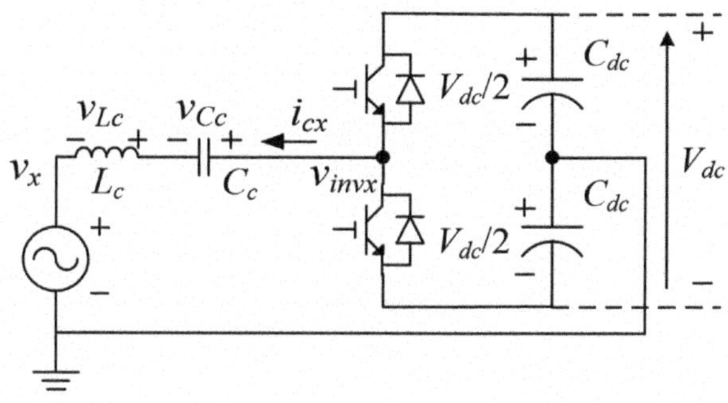

Fig. 2.4 *LC*-HAPF single-phase equivalent circuit

$$v_{invx}(t) = d_x(t)\frac{V_{dc}}{2} = L_{cx}\frac{di_{cx}(t)}{dt} + v_x(t) + v_{cx}(t) \tag{2.2}$$

where d_x refers to the duty ratio in one phase of the VSI in each small-time interval.

For digital implementation of the PWM control algorithm and the consideration about the filter inductance deviation, the instantaneous averaged model in (2.2) can be represented in discrete time with a step size of k as in (2.3):

$$d_x(k)\frac{V_{dc}}{2} = K_d L_{cx}\frac{i_{cx}(k+1) - i_{cx}(k)}{T_s} + v_x(k) + v_{cx}(k) \tag{2.3}$$

where K_d is inductance deviation coefficient that can be treated as control variable, T_s is the sampling period. $i_{cx}(k+1)$ can be substituted by the reference compensating current $i_{cx}{}^{*}$. Hence, the duty ratio of one phase VSI is now obtained as:

$$d_x(k) = \frac{2}{V_{dc}}\left(K_d L_{cx}\frac{i_{cx}{}^{*}(k) - i_{cx}(k)}{T_s} + v_x(k) + v_{cx}(k)\right) \tag{2.4}$$

After that, the duty ratio is used to compare with the fixed frequency triangular carrier wave in order to generate PWM signal for the switching devices in VSI. The switching frequency of the VSI is the same as the frequency of the triangular carrier wave.

2.3.2 *Stability and Parameter Design*

According to (2.4) and Figs. 2.1, 2.2 shows the s-domain closed-loop control block diagram model for the deadbeat current controller of the *LC*-HAPF. The information of each block will be given in below. $G_{PWM}(s)$ is an average s-domain model, note that the PWM VSI can be assumed as unity gain [2]. To accurately describe the $G_{PWM}(s)$, computation and modulation delay in s-domain should be considered, where the computation delay is a unit sampling period T_s and the modulation delay is $0.5T_s$ [3]. The transfer function of the PWM VSI can be expressed as:

$$G_{PWM}(s) = e^{-1.5T_s \cdot s} \tag{2.5}$$

where T_s is the sampling time. For obtaining a rational transfer function, delays are usually approximated by poles and zeros. Then, *Pade*' approximation (2.6) can be applied to (2.5).

$$G_{PWM}(s) = e^{-1.5T_s \cdot s} \approx \frac{1 - 0.75 \cdot T_s \cdot s}{1 + 0.75 \cdot T_s \cdot s} \tag{2.6}$$

In Fig. 2.2, $G_{LC}(s)$ and $G_C(s)$ are the LC coupling branch and coupling capacitor impedance in s-domain, which can be expressed as follows:

$$G_{LC}(s) = \frac{C_c s}{L_c C_c s^2 + 1} \tag{2.7}$$

$$G_C(s) = \frac{1}{C_c s} \tag{2.8}$$

The transfer function according to Fig. 2.2 can be deduced as:

$$\begin{aligned} i_{cx}(s) &= G_{CL\_i}(s) \cdot i_{cx}{}^*(s) + G_{CL\_v}(s) \cdot v_x(s) \\ &= \frac{K_d \frac{L_c}{T_s} G_{PWM}(s) G_{LC}(s)}{1 + K_d \frac{L_c}{T_s} G_{PWM}(s) G_{LC}(s) - G_{PWM}(s) G_{LC}(s) G_C(s)} i_{cx}{}^*(s) \\ &+ \frac{G_{LC}(s)(G_{PWM}(s) - 1)}{1 + K_d \frac{L_c}{T_s} G_{PWM}(s) G_{LC}(s) - G_{PWM}(s) G_{LC}(s) G_C(s)} v_x(s) \end{aligned} \tag{2.9}$$

where $G_{CL\_i}(s)$ is the system closed-loop transfer function between i_{cx} and $i_{cx}{}^*$, $G_{CL\_v}(s)$ is the system closed-loop transfer function between i_{cx} and v_x.

To analyse the stability of the proposed deadbeat current controller, the range of the control coefficient K_d must be found. Theoretically, K_d should be as high as possible to obtain high gain, fast response and low steady-state error at low-order harmonics frequency and fundamental frequency. However, system stability is under the 1st priority consideration. The stability margin would be sacrificed with high K_d gain. Therefore, with the help of the Routh's stability criterion [4] and Fig. 2.2, the analysis of the range of K_d is based on the characteristic equation ($q(s) = 0$) of the closed-loop transfer function between i_{cx} and $i_{cx}{}^*$. The closed-loop transfer function between i_{cx} and $i_{cx}{}^*$ is shown in (2.10) and its characteristic equation can be expressed in (2.11).

$$G_{CL\_i}(s) = \frac{K_d \frac{L_c}{T_s} G_{PWM}(s) G_{LC}(s)}{1 + K_d \frac{L_c}{T_s} G_{PWM}(s) G_{LC}(s) - G_{PWM}(s) G_{LC}(s) G_C(s)} \tag{2.10}$$

$$q(s) = 1 + K_d \frac{L_c}{T_s} G_{PWM}(s) G_{LC}(s) - G_{PWM}(s) G_{LC}(s) G_C(s) = 0 \tag{2.11}$$

According to the stability criterion, the first column of Routh's array needs to be larger than 0 in order to guarantee the system stability. After mathematical calculation, the range of K_d can be obtained as:

$$0 < K_d < \frac{4}{3} + \frac{1}{6} \frac{T_s^2}{L_c C_c} \tag{2.12}$$

Table 2.1 System parameters for MATLAB simulation

System Parameters	Physical Values
L_c, C_c	8 mH, 50 μF
f_1 (fundamental frequency)	50 Hz
f_s (sampling frequency)	10 kHz
f_{sw} (switching frequency: DBCC)	10 kHz
K_d	0.2~1.2

Since the term $1/6 \times T_s^2/L_cC_c$ is very small, it can be neglected for simplicity. Hence, the range of K_d can be obtained as:

$$0 < K_d < \frac{4}{3} \tag{2.13}$$

The deduced model of the deadbeat current control loop of the *LC*-HAPF is analyzed by using MATLAB and the system parameters including different K_d range is given in Table 2.1. As mentioned before, K_d should be selected as high as possible within the bounded value to minimize the steady-state error and shorten the response time. However, high K_d value will cause overshoot problem and lower the stability margin. The bode diagram of the closed-loop current control system $G_{CL\_i}(s)$ is shown in Fig. 2.5, which shows that the system can achieve unity gain with zero phase shifting at the fundamental frequency. That means $i_{cx} \approx i_{cx}^*$, the compensating current can track its reference with less steady-state error, thus obtaining better current tracking performance. To obtain a balance on steady-state error, overshoot and response time, K_d should be selected in around 0.8 to 1.1, which depends on desired performance. In this work, $K_d = 0.8$ is chosen.

The bode diagram of the closed-loop transfer function $G_{CL\_v}(s)$ between i_{cx} and v_x with $K_d = 0.8$ is shown in Fig. 2.6. It indicates that the proposed deadbeat controller provides sufficient attenuation to the disturbance from the grid-side voltage v_x within the closed-loop bandwidth. Note that the glitches on 250 Hz are the resonant frequency (250 Hz) of the *LC* branch. It is designed to suppress the 5th order harmonics.

2.4 Overall Control Block Diagram

The overall control block diagram of the proposed deadbeat controller for the *LC*-HAPF is shown in Fig. 2.7. The reference compensating currents for the *LC*-HAPF (i_{cx}^*, the subscript $x = a, b, c$ for three-phases) are determined based on the three-phase instantaneous *p*-*q* theory [1]. Then the load voltage v_x, coupling capacitor voltage v_{cx}, compensating current error Δi_{cx}, controller variable K_d, sampling time T_s, coupling inductance L_{cx}, will be sent to the deadbeat current PWM control part

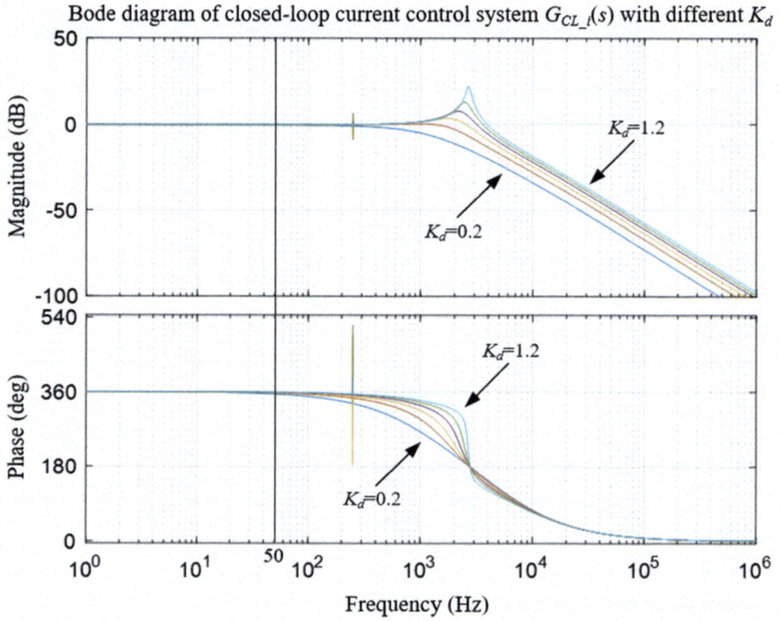

Fig. 2.5 Bode diagram of closed-loop current control system $G_{CL\_i}(s)$ with different K_d varies from 0.2 to 1.2

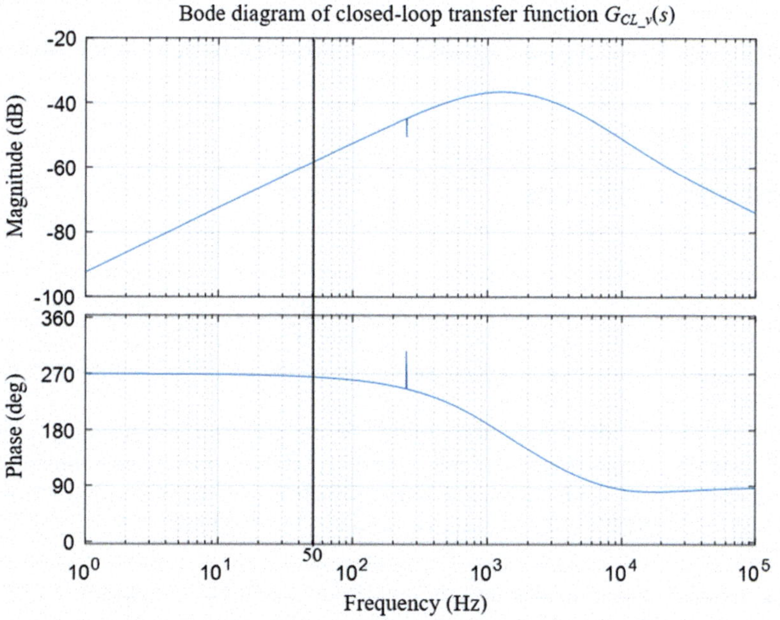

Fig. 2.6 Bode diagram of closed-loop transfer function $G_{CL\_v}(s)$ with $K_d = 0.8$

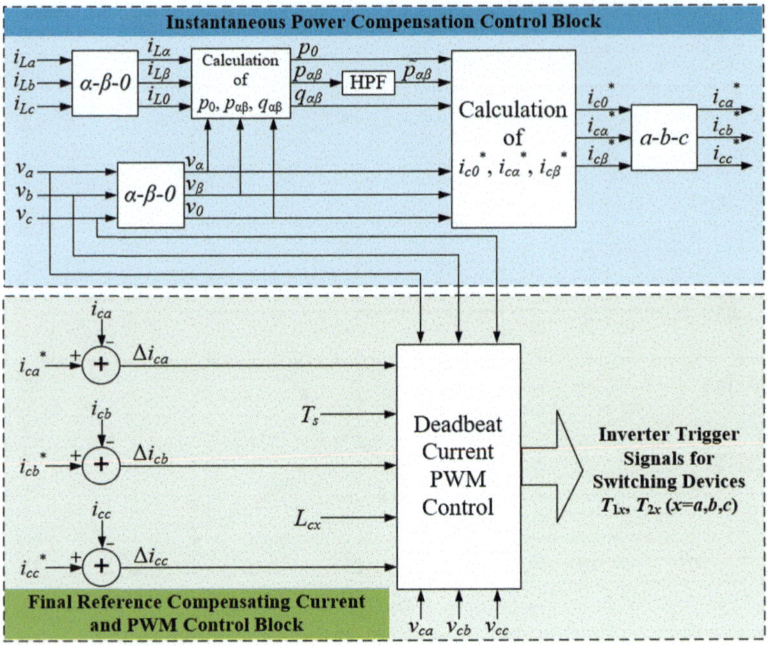

Fig. 2.7 Overall control block diagram of *LC*-HAPF for DBCC

to generate the PWM trigger signals for the switching devices of the VSI of the *LC*-HAPF.

2.5 Simulation Results

In this section, simulation results will be presented to verify the proposed deadbeat current controller of the *LC*-HAPF under three-phase power quality compensator application. The system parameters for simulations are given in Table 2.2. Note that the settling time T_s are obtained after switching on the *LC*-HAPF with loads already connected.

In the following, the simulation studies of the conventional hysteresis current control (HCC), proportional-integral control (PIC), proportional-multi-resonant control (PMRC) and deadbeat current control (DBCC) for the three-phase four-wire center-split *LC*-HAPF are carried out by using PSCAD/EMTDC. The corresponding simulation results before and after the hysteresis current controlled, PI controlled, MPR controlled and deadbeat current controlled *LC*-HAPF compensation will be summarized in Figs. 2.8, 2.9, 2.10 and Table 2.3. With reference to the IEEE standard 519–2014 [5], the acceptable Total Demand Distortion (TDD) $\leq 15\%$ with I_{SC}/I_L is in 100 < 1000 scale (a small rating 110V-5kVA system). The nominal rate

Table 2.2 System parameters for simulation and experiment

System parameters	Physical values
v_{gx}, f_g	110 V_{rms}, 50 Hz
L_s	0.1 mH
L_c, C_c	8 mH, 50 μF
C_{dc}, V_{dc}	1000 μF, 80 V
L_{NL}, C_{NL}, R_{NL}	35 mH, 400 μF, 43 Ω
f_S (sampling frequency)	10 kHz
f_{sw} (switching frequency: DBCC)	10 kHz
K_d	0.8

current is assumed to be equal to the fundamental load current at the worst-case analysis, which results in Total Harmonic Distortion (THD) = TDD $\leq$ 15%. Therefore, this book evaluates the *LC*-HAPF current harmonics compensating performance by setting an acceptable THD $\leq$ 15%.

The basic control principle of the HCC is based on deriving the switching signals from the comparison of the current error Δi_{cx} between the reference compensating current i_{cx}^* and the actual compensating current i_{cx} with a fixed hysteresis band H. The trigger signals will drive the switching devices of the VSI in order to let i_{cx} tracks with its reference i_{cx}^*. Every time when the current error Δi_{cx} is larger than the positive or smaller than the negative hysteresis band's boundary $i_{cx}^* + H$ or i_{cx}^*-H, a state change of the VSI's trigger signals occurs [6].

The PIC and PMRC control are initially computing the current error signal Δi_{cx} by subtracting the reference compensating current i_{cx}^* by the actual compensating current i_{cx}, Δi_{cx} will be inputted into the PIC / PMRC. Then its corresponding output will be compared with a fixed frequency triangular carrier wave V_{tri} in order to generate the trigger signals for the switching devices of the VSI [3].

(A) Hysteresis Current Control: Fig. 2.8a shows the simulation results of HCC for the *LC*-HAPF. From Fig. 2.8a, the system current ripples after compensation are obvious. From Fig. 2.8a and Table 2.3, although it can compensate the system power factor (PF) from 0.8 to 0.98, the system neutral current from 3.08Arms to 0.75Arms, the total current harmonics distortion (THDi_{sx}) from 35% to about 11%, which satisfies the international standard of THD $\leq$ 15% [5]. The error percentage (Q_{error}) of reactive power injection is 1.9%. Figure 2.9a also illustrates the phase *a* source current harmonics spectrum after compensation. Figure 2.10a clearly illustrates the switching frequency of HCC is not fixed. The noise and distortion mainly focus on low frequency band (highlighted in Fig. 2.10a with dotted square) as the VSI is switching with a low frequency.

(B) Proportional-Integral Control: Fig. 2.8b shows the simulation results of PIC for the *LC*-HAPF. The system current ripples after compensation have been reduced significantly compared with the HCC. From Fig. 2.8b and Table 2.3, it can compensate the system PF from 0.8 to 0.99, the system neutral current from 3.08 Arms to 0.60 Arms, the THDi_{sx} from 35% to about 8%, which satisfies the

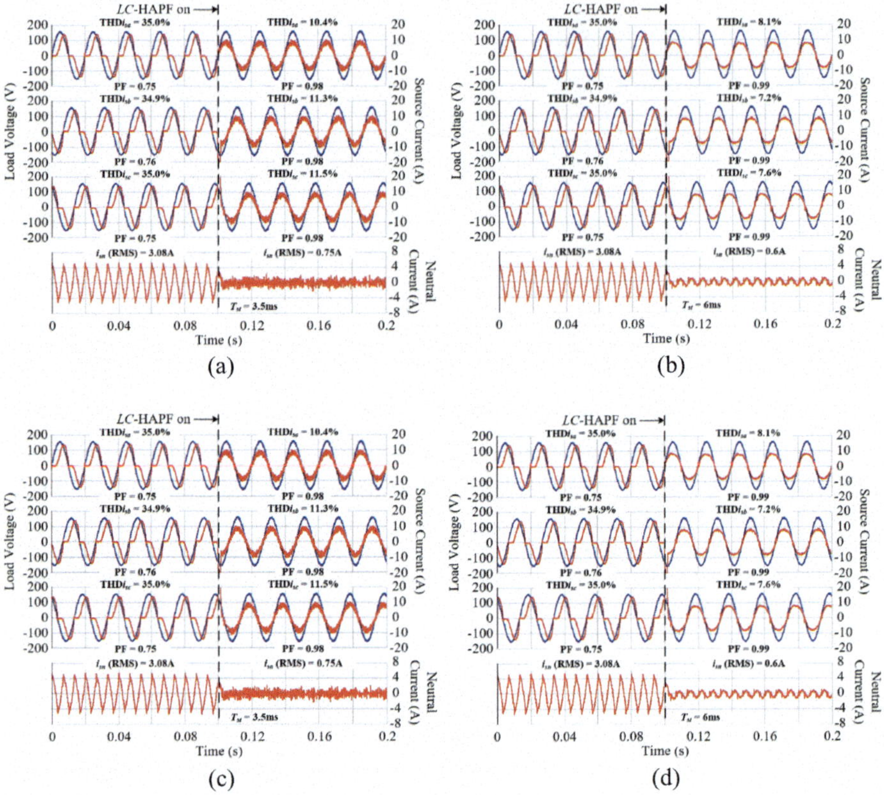

Fig. 2.8 Simulated waveforms before and after *LC*-HAPF compensation with different control strategies: **a** HCC **b** PIC **c** PMRC **d** DBCC

Table 2.3 Simulation results of HCC, PIC, PMRC and DBCC for *LC*-HAPF before and after compensation

	P_{sx} (W)	Q_{sx} (var)	PF	i_{sx} (A)	i_{sn} (A)	Worst THDi_{sx} (%)	T_{st} (ms)
Before Compensation	271.6	203.9	0.75	3.28	3.08	35.0	/
HCC	271.8	4.0	0.98	2.68	0.75	11.5	3.5
PIC	270.5	2.5	0.99	2.47	0.60	8.1	6.0
PMRC	270.5	2.0	0.99	2.46	0.46	6.8	50.0
DBCC	271.0	1.8	0.99	2.46	0.40	6.4	3.5

international standard of THD $\leq$ 15% [5]. Figure 2.9b also illustrates the phase *a* source current harmonics spectrum after compensation. Since the PIC uses carrier-based PWM technique to generate trigger signals for the VSI, the switching frequency

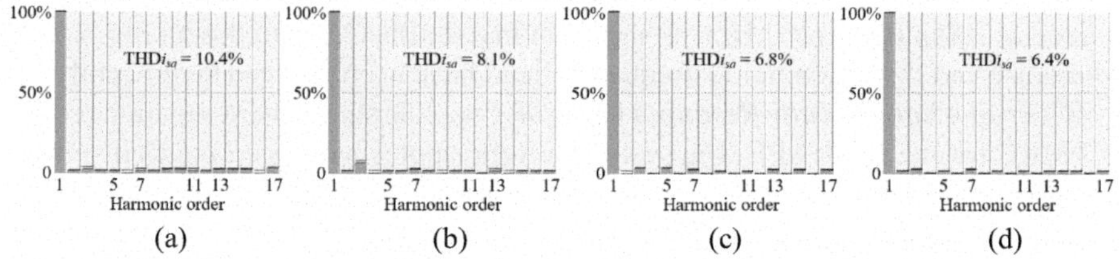

Fig. 2.9 Simulated source current harmonics spectrum (phase a) before and after LC-HAPF compensation with different control strategies: **a** HCC **b** PIC **c** PMRC **d** DBCC

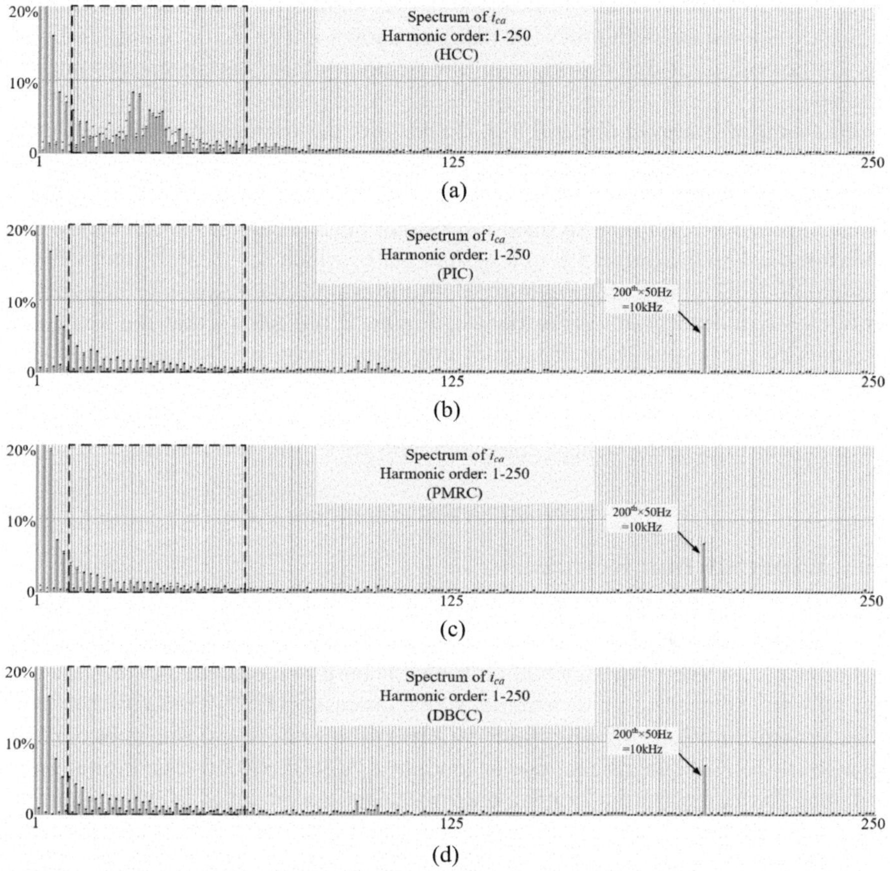

Fig. 2.10 Simulation results of compensating current i_{cx} harmonics spectrum (phase a) with different control strategies: **a** HCC **b** PIC **c** PMRC **d** DBCC

is fixed at 200th order (10 kHz), as shown in Fig. 2.10b. Compared with Fig. 2.10a, both noise and distortion in low frequency band have been reduced significantly.

(C) Proportional-Multi-Resonant Control: Fig. 2.8c shows the simulation results of PMRC for the *LC*-HAPF. The system current ripples have been reduced significantly compared with the HCC after compensation. From Fig. 2.8c and Table 2.3, it can compensate the system PF from 0.8 to 0.99, the system neutral current from 3.08Arms to 0.46Arms, the THDi_{sx} from 35% to less than 7%, which satisfies the international standard of THD $\leq$ 15% [5]. Figure 2.9c also illustrates the phase a source current harmonics spectrum after compensation. Since the PMRC uses carrier-based PWM technique to generate trigger signals for the VSI, the switching frequency is fixed at 200th order (10 kHz), as shown in Fig. 2.10c. Compared with Fig. 2.10a, both noise and distortion in low frequency band have been significantly reduced.

(D) Deadbeat Current Control: Fig. 2.8d shows the simulation results of DBCC for the *LC*-HAPF. Compared with HCC case, the system current ripples after compensation have been reduced significantly. From Fig. 2.8d and Table 2.3, it can compensate the system PF from 0.8 to 0.99, the system neutral current from 3.08Arms to 0.4Arms, the THDi_{sx} from 35% to less than 6.5%, which satisfies the international standard of THD $\leq$ 15% [5]. Figure 2.9d also illustrates the phase a source current harmonics spectrum after compensation. Figure 2.10d shows that the switching frequency of the proposed deadbeat current control is fixed at 200th order (10 kHz). Compared with Fig. 2.10a, both noise and distortion in low frequency band have been reduced significantly. Therefore, the proposed deadbeat current controller obtains the best compensation performances compared with HCC, PIC and PMRC.

2.6 Experimental Results

For verifying the simulation results, a small rating 110V-5kVA three-phase four-wire center-split *LC*-HAPF experimental prototype is built in the laboratory as shown in Fig. 2.11. Its' control system is a digital signal processor (DSP) TMS320F2407 and the sampling time T_s is 0.1 ms ($f_S = 10$ kHz). The switching devices for the VSI are Mitsubishi IGBT intelligent power modules PM300DSA060. The limitation of switching frequency for the switching devices is at maximum of 20 kHz while the switching frequency f_{sw} of the deadbeat current controlled *LC*-HAPF experiment is 10 kHz. The corresponding experimental results before and after the hysteresis current controlled, PI controlled, PMR controlled and deadbeat current controlled *LC*-HAPF compensation are summarized in Figs. 2.12, 2.13, 2.14 and Table 2.4. The system parameters for experiment are given in Table 2.2.

***(A) Hysteresis Current Control*:** Fig. 2.12a shows the experimental results of HCC for the *LC*-HAPF. From Fig. 2.12a, the system current ripples after compensation are obvious, which is similar to the simulation. From Fig. 2.12a and Table 2.4, it can compensate the system PF from 0.84 to 0.98, the system neutral current from 2.6Arms to 0.9Arms, the THDi_{sx} from 29% to about 12%, which satisfies

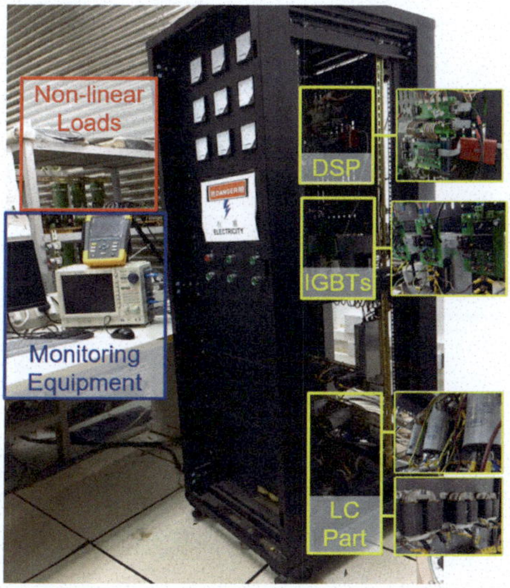

Fig. 2.11 Experimental prototype of 110V-5kVA *LC*-HAPF

the international standard of THD $\leq$ 15% [5]. The transient response is the fastest among the controllers, which is 6 ms. Figure 2.13a also illustrates the source current harmonics spectrum after compensation. Figure 2.14a clearly illustrates that the switching frequency of HCC is not fixed. The noise and distortion mainly focus on lower frequency band (highlighted in Fig. 2.14a with dotted square) as the VSI is switching with a low frequency.

(B) Proportional-Integral Control**:** Fig. 2.12b shows the experimental results of PIC for the *LC*-HAPF. Compared with HCC, the system current ripples after compensation have been significantly reduced. From Fig. 2.12b and Table 2.4, it can compensate the system PF from 0.8 to 0.99, the system neutral current from 2.6Arms to 0.6Arms, the THDi_{sx} from 29% to about 11%, which satisfies the international standard of THD $\leq$ 15% [5]. The transient response is a bit slower than HCC and DBCC because of the integral action. Figure 2.13b also illustrates the phase *a* source current harmonics spectrum after compensation. The switching frequency of PIC is fixed because it is also used carrier-based PWM technique to generate trigger signals for the VSI. Figure 2.14b shows that the switching frequency is fixed at 200th order (10 kHz) and the distortion and noise at low frequency band has been reduced compared to the HCC in Fig. 2.14a.

(C) Proportional-Multi-Resonant Control**:** Fig. 2.12c shows the experimental results of PMRC for the *LC*-HAPF. Compared with the HCC, the system current ripples after compensation have been significantly reduced. From Fig. 2.12c and Table 2.4, it can compensate the system PF from 0.8 to 0.99, the system neutral current from 2.6Arms to 0.6Arms, the THDi_{sx} from 29% to about 10%, which satisfies the international standard of THD $\leq$ 15% [5]. The transient response is the slowest among the controllers because of the integral action. Figure 2.13c also illustrates

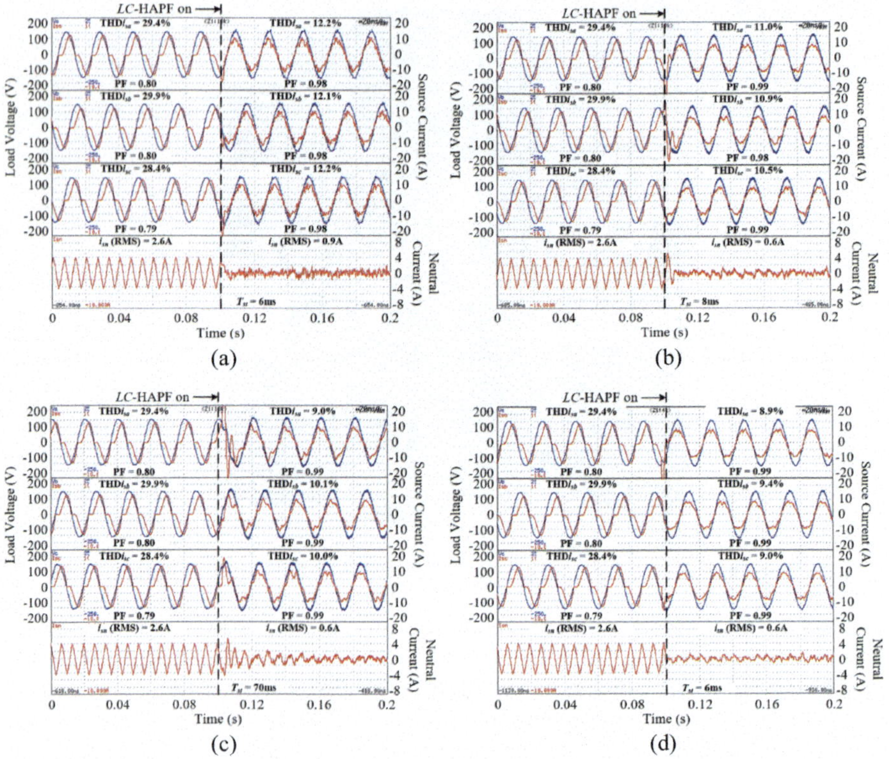

Fig. 2.12 Experimental waveforms before and after *LC*-HAPF compensation with different control strategies: **a** HCC **b** PIC **c** PMRC **d** DBCC

Table 2.4 Experimental results of HCC, PIC, PMRC and DBCC for *LC*-HAPF before and after compensation

	P_{sx} (W)	Q_{sx} (var)	PF	i_{sx} (A)	i_{sn} (A)	Worst THDi_{sx} (%)	T_{st} (ms)
Before Compensation	290	190	0.80	3.3	2.6	29.9	/
HCC	310	34	0.98	3.0	0.9	12.2	6
PIC	300	17	0.99	2.7	0.6	11.0	8
PMRC	300	9	0.99	2.7	0.6	10.0	70
DBCC	300	17	0.99	2.7	0.6	9.4	6

the phase *a* source current harmonics spectrum after compensation. The switching frequency of PMRC is fixed because it is also used carrier-based PWM technique to generate trigger signals for the VSI. Figure 2.14c shows that the switching frequency is fixed at 200th order (10 kHz) and the distortion and noise at low frequency band has been reduced compared to the HCC in Fig. 2.14a.

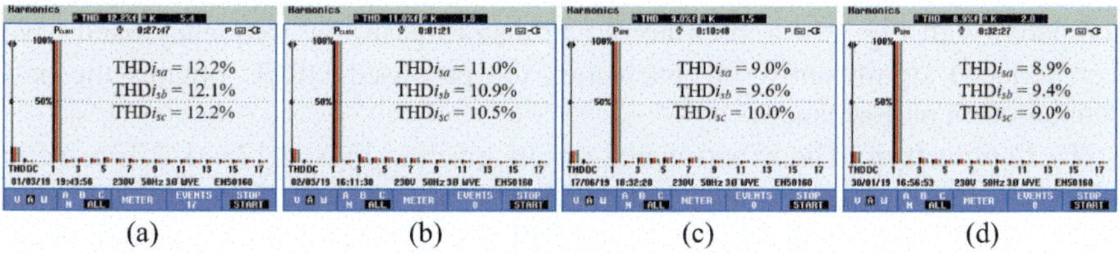

Fig. 2.13 Experimental source current harmonics spectrum (phase *a*) before and after *LC*-HAPF compensation with different control strategies: **a** HCC **b** PIC **c** PMRC **d** DBCC

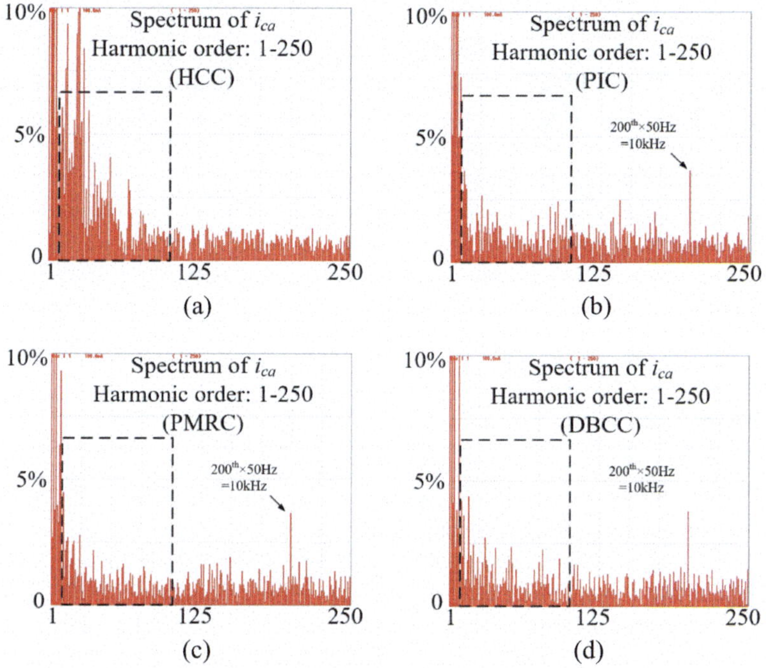

Fig. 2.14 Experimental results of compensating current i_{cx} harmonics spectrum (phase *a*) with different control strategies: **a** HCC **b** PIC **c** PMRC **d** DBCC

(D) Deadbeat Current Control: Fig. 2.12d shows the experimental results of DBCC for the *LC*-HAPF. Compared with the HCC case, the system current ripples after compensation have been reduced significantly. From Fig. 2.12d and Table 2.4, it can compensate the system PF from 0.84 to 0.99, the system neutral current from 2.6Arms to 0.6Arms, the THDi_{sx} from 29% to less than 10%, which satisfies the international standard of THD $\leq$ 15% [5]. The transient response is 6 ms (as fast as HCC) because of the deadbeat characteristic. Figure 2.13d also illustrates the source current harmonics spectrum after compensation. Figure 2.14d shows that the switching frequency of the proposed DBCC is fixed at 200th order (10 kHz).

Compared with Fig. 2.14a, both noise and distortion in low frequency band have been reduced significantly. In conclusion, the proposed DBCC obtains the best compensation performances.

***(E) Comparison*:** The experimental results given in Figs. 2.12a–d, 2.13a–e and Table 2.4 are consistent with the simulation results in Figs. 2.8a–d, 2.9a–d and Table 2.3. This verifies that the proposed DBCC can be applied for the three-phase four-wire *LC*-HAPF performing power quality compensation. In addition, the HCC, PIC, PMRC and DBCC for the *LC*-HAPF can compensate the system PF to 0.98 or above. And the DBCC can obtain the best THDi_{sx} and neutral current compensation. Figures 2.9a–d and 2.13a–b shows the simulated and experimental harmonics spectrum of source current with the HCC, PIC, PMRC and DBCC. Obviously, the proposed DBCC can suppress harmonic components more effectively. Compared with the HCC, PIC and PMRC, the DBCC obtains better compensation performances with low output current ripple (noise), fixed switching frequency, lowest steady-state error and better neutral current and current harmonics compensation. Moreover, it can provide fast dynamic response.

Comparing the results between simulation and experiment sections, the differences between them are caused by the noise in the experiment equipment, the digital computation error, inaccurate component parameters and resolution of the transducers.

2.7 Summary

A DBCC strategy of *LC*-HAPF is proposed in this chapter, which can provide a cost-effective power quality compensation solution with small steady-state error, fast dynamic response and low output current ripples. First of all, the mathematical modeling and the control block diagram of the proposed deadbeat current controlled *LC*-HAPF is derived and built. Based on it, the stability issue and parameter design consideration are analyzed and discussed. Finally, representative simulation and experimental results of the DBCC for the *LC*-HAPF are given in comparison with conventional HCC, PIC and PMRC, which shows its effectiveness and superior compensating performances.

References

1. H. Akagi, Y. Kanazawa, A. Nabae, Instantaneous reactive power compensators comprising of switching devices without energy storage components," IEEE Trans. Ind. Appl., **20**(3), (1984). pp. 625 630
2. T. Ye, N.-Y Dai, C.-S. Lam, M.-C. Wong, J.M. Guerrero, Analysis, design, and implementation of a quasi proportional-resonant controller for a multifunctional capacitive-coupling grid-connected inverter, IEEE Trans. Ind. Appl., **52**(5), (2016) pp. 4269–4280
3. S. Buso, P. Mattavelli, Digital control in power electronics. Morgan & Claypool, (2015)

4. R.C. Dorf, R.H. Bishop, Modern control systems, 12th ed. Pearson Education, (2010)
5. IEEE Standard 519–2014: IEEE recommended practices and requirements for harmonic control in electrical power systems. (2014)
6. C.-S. Lam, M.-C. Wong, Y.D. Han, Hysteresis current control of hybrid active power filters. IET Power Electron. **5**(7), 1175–1187 (2012)

Chapter 3
Improved Model Predictive Control for *LC*-HAPF

3.1 Introduction

In this chapter, an improved model predictive control (MPC) with signal correction technique for an *LC*-coupling hybrid active power filter (*LC*-HAPF) is proposed, which can track the reference compensating current with fast dynamic response and low steady-state error. The conventional MPC causes certain degree of steady-state error due to the error from discretization and prediction. To solve this problem, a signal correction technique is proposed to improve the compensation performance of the MPC for the LC-HAPF. First of all, the model of the *LC*-HAPF for implementing the MPC is deduced in this chapter. Then, the control objective and the design methodology of the improved MPC are presented. After that, the stability and parameter design of the improved MPC is analyzed and discussed. Finally, the compensation performance of the proposed improved MPC for the *LC*-HAPF is verified by simulation and experimental results in comparison with the conventional MPC and other representative current controllers.

The main contributions of this chapter are summarized in below:

(1) Derive the system model of the *LC*-HAPF for implementing the MPC.
(2) Propose an improved MPC for the *LC*-HAPF, which provides low steady-state error and fast dynamic response performances.
(3) Study the stability and the parameter design of the improved MPC for the *LC*-HAPF to ensure its compensation performance.
(4) Verify the effectiveness of the improved MPC for the *LC*-HAPF by simulation and experimental results in comparison with the HCC, QPRC and the conventional MPC.

C.-S. Lam et al., *Advanced Controllers Design for Hybrid Active Power Filters*,
https://doi.org/10.1007/978-981-97-6847-9_3

3.2 Design of Improved Model Predictive Control for *LC*-HAPF

3.2.1 Control Objective

The main idea of the conventional MPC is using the sensed signals, the system parameters, and the reference compensating current from *p-q* theory [1] to predict compensating current, so as to minimize the compensating current tracking error and improve the dynamic response. That is, all possible predicted compensating current $i_{cx}(k+1)$ in each phase is calculated by using the predictive model. After the rolling optimization, the optimal $i_{cx}(k+1)$ that can minimize the cost function J_x will be determined. Hence, the corresponding optimal switching state S_x and the trigger signals T_{1x} and T_{2x} for all the switching devices will be found and applied to the voltage source inverter (VSI) of the *LC*-HAPF. The details of the optimization process are provided in this section.

3.2.2 Conventional Model Predictive Control for LC-HAPF

From Fig. 1.4, the single-phase equivalent circuit of an *LC*-HAPF system is illustrated in Fig. 3.1, where v_{cx} represents the coupling capacitor voltage. According to this, the differential equation of the *LC*-HAPF in each phase can be expressed as:

$$v_{invx}(t) = L_c \frac{di_{cx}(t)}{dt} + v_x(t) + v_{cx}(t). \tag{3.1}$$

For getting the discrete model, the time derivative of compensating current $di_{cx}(t)/dt$ can be discretized and approximated in (3.2) by using the Forward Euler Method

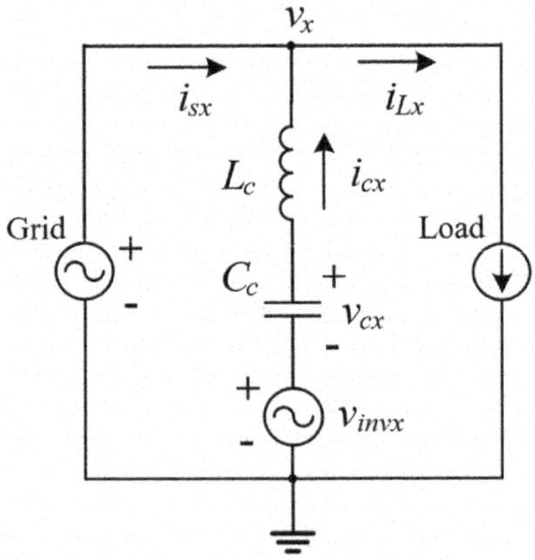

Fig. 3.1 Single-phase equivalent circuit of *LC*-HAPF

[2].

$$\frac{di_{cx}(t)}{dt} \approx \frac{i_{cx}(k+1) - i_{cx}(k)}{T_s}. \tag{3.2}$$

From (3.1) and (3.2), the predicted compensating current $i_{cx}(k+1)$ in discrete-time domain is expressed as:

$$i_{cx}(k+1) = i_{cx}(k) + \frac{T_s}{L_c}\left[v_{invx}(k) - v_x(k) - v_{cx}(k)\right] \tag{3.3}$$

where $v_{invx}(k)$ is the optimal inverter voltage to be evaluated. The relationship between the inverter voltage and the corresponding switching states S_x of each phase is listed in Table 3.1. Furthermore, S_x in each phase can be expressed as (3.4).

$$S_x(k) = \begin{cases} 0, & T_{1x} \text{ off, } T_{2x} \text{ on;} \\ 1, & T_{1x} \text{ on, } T_{2x} \text{ off;} \end{cases}. \tag{3.4}$$

Due to the three-phase-four-wire system characteristic, the optimal S_x can be found in each phase by the independent cost function J_x. Based on [3], the cost function J_x for the conventional MPC is chosen as:

$$J_x(k) = \left\| i_{cx}{}^*(k+1) - i_{cx}(k+1) \right\|^2 \tag{3.5}$$

To obtain $i_{cx}{}^*(k+1)$, Lagrange extrapolation method is used. Due to the constant sampling period, Lagrange extrapolation performs linear prediction to estimate $i_{cx}{}^*(k+1)$ [3]. And the second order Lagrange extrapolation is used to ensure the system simplicity and prediction accuracy. The prediction of the reference compensating current is expressed as:

$$i_{cx}{}^*(k+1) \approx 2i_{cx}{}^*(k) - i_{cx}{}^*(k-1) \tag{3.6}$$

where $i_{cx}{}^*$(k) can be obtained by using the instantaneous p-q theory [1]. By substituting the $i_{cx}(k+1)$ and $i_{cx}{}^*(k+1)$ in to (3.5) with the help of (3.3) and (3.6), the optimal $i_{cx}(k+1)$ that can minimize J_x will be determined and the optimal S_x will be applied to control the VSI of the *LC*-HAPF. In addition, the control block diagram of the conventional MPC is given in Fig. 3.2.

Table 3.1 Switching state of each phase

Switching state S_x	Upper trigger signals T_{1x}	Lower trigger signals T_{2x}	Inverter voltage v_{invx}
0	0	1	$-V_{dc}/2$
1	1	0	$V_{dc}/2$

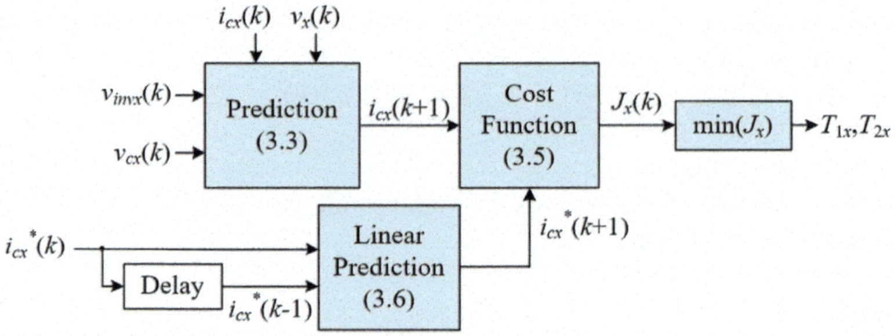

Fig. 3.2 Control block diagram of conventional MPC

3.2.3 *Proposed Model Predictive Control for* LC-*HAPF*

The discretization and prediction used in the conventional MPC enlarge the steady-state error [4, 5]. Therefore, this work also proposes a signal correction technique to address this issue and improve the compensation performance through reducing the steady-state error. The design process of the proposed MPC for *LC*-HAPF is elaborated in below.

From (3.1) and (3.2), the reference inverter voltage $v_{invx}{}^{*}(k)$ in discrete-time domain is expressed as:

$$v_{invx}{}^{*}(k) = \frac{L_c}{T_s}[i_{cx}(k+1) - i_{cx}(k)] + v_x(k) + v_{cx}(k). \tag{3.7}$$

In the MPC, the optimization is done based on $v_{invx}{}^{*}$ calculated in each phase. For achieving better optimization, the $v_{invx}{}^{*}(k)$ should be as ideal as possible. Therefore, the compensating current in the next cycle $i_{cx}(k+1)$ should be replaced by its ideal value, which is the reference compensating current in the next cycle $i_{cx}{}^{*}(k+1)$ as shown in (3.8) and it can be obtained from (3.6).

$$i_{cx}(k+1) = i_{cx}{}^{*}(k+1) \tag{3.8}$$

Then, via (3.6)- (3.8), the $v_{invx}{}^{*}(k)$ for the conventional MPC is expressed as:

$$v_{invx}{}^{*}(k) = \frac{L_c}{T_s}\left[2i_{cx}{}^{*}(k) - i_{cx}{}^{*}(k-1) - i_{cx}(k)\right] + v_x(k) + v_{cx}(k) \tag{3.9}$$

For reducing the error produced by the discretization in (3.2) and the reference compensating current prediction in (3.6), a signal correction technique [6] is applied in this work. Hence, based on (3.9), the modified reference inverter voltage is expressed as:

Fig. 3.3 Optimization algorithm for each phase

```
Initialization: S_x = 0;
Calculate v_invx* by (3.10)
while S_x ≤ 1:
        Find v_invx[S_x] by Table 3.1
        Calculate J_x[S_x] by (3.11)
if min(J_x) == J_x[0]:
        T_1x = 0; T_2x = 1;
else if min(J_x) == J_x[1]:
        T_1x = 1; T_2x = 0;
S_x++;
```

$$v_{invx}{}^{*}(k) = \frac{L_C}{T_s}\left[2i_{cx}{}^{*}(k) - i_{cx}{}^{*}(k-1) - i_{cx}(k)\right] + v_x(k) + v_{Cx}(k) + K_e\frac{L_C}{T_s}\left[i_{cx}{}^{*}(k) - i_{cx}(k)\right] \tag{3.10}$$

where K_e is a control coefficient that determines the influence of the compensating current error on the reference inverter voltage. The design of K_e will be discussed in Sect. 3.2.4.

Since the optimized target changes from $i_{cx}(k+1)$ in the conventional MPC to the $v_{invx}{}^{*}(k)$ in the proposed MPC, the cost function is changed as (3.11), in which the $v_{invx}(k)$ can be found in Table 3.1. After the rolling optimization, the optimal $v_{invx}(k)$ that can minimize J_x will be determined and the optimal S_x will be applied to control the VSI of the *LC*-HAPF.

$$J_x(k) = \left\| v_{invx}{}^{*}(k) - v_{invx}(k) \right\|^2 \tag{3.11}$$

In addition, the optimization algorithm of the proposed method is given in Fig. 3.3, which is written in pseudocode.

3.2.4 *Stability and Parameter Design of Model Predictive Control*

Routh-Hurwitz stability criterion [7] and Nyquist stability criterion [8] cannot be applied for the stability analysis for the MPC controlled *LC*-HAPF because the MPC is a nonlinear control strategy. Therefore, Lyapunov stability criterion is used to discover how K_e affects the stability as it can be applied in the non-linear system [9].

The Lyapunov stability criterion is directly applied to the compensating current error Δi_{cx} so as to determine the stability of the whole system. From [10], (3.10) is exponentially stable if there exists a Lyapunov function $F[\Delta i_{cx}(k)]$ which satisfies the following inequalities:

$$\begin{aligned}
&F[\Delta i_{cx}(k)] \geq c_1|\Delta i_{cx}(k)|^l, \quad \forall \Delta i_{cx}(k) \in C;\\
&F[\Delta i_{cx}(k)] \leq c_2|\Delta i_{cx}(k)|^l, \quad \forall \Delta i_{cx}(k) \in D;\\
&\Delta F[\Delta i_{cx}(k)] = F[\Delta i_{cx}(k+1)] - F[\Delta i_{cx}(k)] < -c_3|\Delta i_{cx}(k)|^l + c_4
\end{aligned} \tag{3.12}$$

where $l \geq 1$, cm (m = 1, 2, 3, 4) is a positive constant, $C \subseteq \mathbb{R}$ is a positive control invariant set and $D \subseteq C$ is a compact set. According to the Lyapunov's second method for stability, the Lyapunov function $F[\Delta i_{cx}(k)]$ must be a positive-definite function and it is defined in the following:

$$F[\Delta i_{cx}(k)] = \frac{1}{2}\Delta i_{cx}{}^2(k) \tag{3.13}$$

In addition, the rate of change of (3.13) is expressed as:

$$\Delta F[\Delta i_{cx}(k)] = F[\Delta i_{cx}(k+1)] - F[\Delta i_{cx}(k)] \tag{3.14}$$

Based on (3.6) and (3.10, 3.14) becomes:

$$\begin{aligned}
\Delta F[\Delta i_{cx}(k)] = \frac{1}{2}\{&i_{cx}{}^*(k+1) - i_{cx}{}^*(k+2) - \frac{T_s}{L_c}[v_x(k+1) + v_{cx}(k+1) - v_{invx}(k+1)]\\
&+K_e\Delta i_{cx}(k+1)\}^2 - \frac{1}{2}\Delta i_{cx}{}^2(k)
\end{aligned} \tag{3.15}$$

Considering that the parameters of grid and load are bounded, both v_x(k) and $i_{cx}{}^*$(k) should be bounded [11] for all k. From Table 3.1, $v_{invx}(k+1)$ is bounded by $|v_{invx}(k+1)| \leq V_{dc}/2$. The boundary of $v_{cx}(k+1)$ is defined as γ, i.e., $\gamma \geq |v_{cx}(k+1)|$. The maximum voltage drop of the coupling capacitor voltage V_{cx} is calculated as $V_{cx} = V_x + V_{Lcx} + V_{dc}/2$, where $V_{Lcx} = 2\pi f L_c d$ is the voltage drop of the coupling inductor and d is given in below paragraph. According to Table 3.2, the boundary of $v_{cx}(k+1)$, γ is considered as 215.5 V. Then, (3.15) becomes:

$$\Delta F[\Delta i_{cx}(k)] \leq -\frac{1}{2}\Delta i_{cx}{}^2(k) + \frac{1}{2}\left[\frac{T_s}{L_c}\left(\gamma - \frac{V_{dc}}{2}\right) + K_e\Delta i_{cx}(k+1)\right]^2 \tag{3.16}$$

The system will be stable if $\Delta i_{cx}(k+1) \leq \Delta i_{cx}(k)$. Hence, the boundary of the compact set D is defined as d, i.e., $d \geq |\Delta i_{cx}(k)|$. Before the *LC*-HAPF is switched on, the maximum $\Delta i_{cx}(k)$ is equal to the peak value of $i_{cx}{}^*(k)$. In order to reserve some margin, d can be set as a little bit larger than the peak value of the $i_{cx}{}^*(k)$. Considering the peak value of the $i_{cx}{}^*(k)$ for two testing loads, d is chosen as 3.8A. Then, (3.16) becomes:

Table 3.2 System parameters for simulation and experiment

System parameters	Physical values
v_{gx}, f_g	110 V_{rms}, 50 Hz
L_s	0.5 mH
L_c, C_c	8 mH, 50 μF
C_{dc}, V_{dc}	1000 μF, 100 V
L_{NL}, C_{NL}, R_{NL} (nonlinear rectifier load)	35 mH, 400 μF, 43 Ω
L_{LL}, R_{LL} (linear load)	70 mH, 60 Ω
f_s, T_s	10 kHz, 0.1 ms
f_{cb}, B, f_{ch}	50 Hz, 7 Hz, 5 Hz
H	0.156
K_p, K_r, ω_c, ω_1	50, 7150, 6.28 rad/s, 314 rad/s
K_e	0.7

$$\Delta F[\Delta i_{cx}(k)] \le \frac{1}{2}\left[K_e^2 - 1\right]\Delta i_{cx}{}^{*}(k) + \frac{1}{2}\left(\frac{T_s}{L_c}\right)^2\left(\gamma - \frac{V_{dc}}{2}\right)^2 + K_e\frac{T_s}{L_c}\left(\gamma - \frac{V_{dc}}{2}\right)d \tag{3.17}$$

Then, the inequality in (3.12) can be satisfied by setting:

$$\begin{aligned} c_1 &= c_2 = 1; \\ c_3 &= \frac{1}{2}\left[K_e^2 - 1\right]; \\ c_4 &= \frac{1}{2}\left(\frac{T_s}{L_c}\right)^2\left(\gamma - \frac{V_{dc}}{2}\right)^2 + K_e\frac{T_s}{L_c}\left(\gamma - \frac{V_{dc}}{2}\right)d \end{aligned} \tag{3.18}$$

Lastly, the inequality about $\Delta F[\Delta i_{cx}(k)]$ from (3.12) can be expressed as:

$$\Delta F[\Delta i_{cx}(k)] \le -2c_3 F[\Delta i_{cx}(k)] + c_4 \tag{3.19}$$

The inequality (3.19) demonstrates that, $\Delta i_{cx}(k)$ converges in the compact set along with the time as shown in the following:

$$E = \left\{ \begin{aligned} &\Delta i_{cx}(k) \,|\, \|\Delta i_{cx}(k)\| \le \sqrt{\frac{c_4}{c_3}} \\ &= \sqrt{\frac{T_s^2\left(\gamma - \frac{V_{dc}}{2}\right)^2 + 2K_e L_c T_s\left(\gamma - \frac{V_{dc}}{2}\right)d}{K_e^2 L_c^2 - 1}} \end{aligned} \right\} \tag{3.20}$$

By (3.20), define the boundary of compact set E as e in order to obtain the boundary function $e(K_e)$ in (3.21), where K_e is considered as the only variable.

$$e(K_e)=\sqrt{\frac{T_s^2\left(\gamma-\frac{V_{dc}}{2}\right)^2+2K_eL_cT_s\left(\gamma-\frac{V_{dc}}{2}\right)d}{K_e^2L_c^2-1}} \tag{3.21}$$

Based on (3.21), $e(K_e)$ is defined if and only if the inequality in (3.22) is satisfied.

$$K_e\geq\frac{T_s\left(V_{dc}^2+4\gamma V_{dc}-4\gamma^2\right)}{4L_c(2\gamma-V_{dc})d} \tag{3.22}$$

If K_e satisfies (3.22), the boundary $e(K_e)$ exists, which produces the boundary of $\Delta i_{cx}(k)$ according to (3.20) and (3.21). Moreover, the system (3.10) is stable if the boundary of $\Delta i_{cx}(k)$ exists. Therefore, the system (3.10) is stable if the inequality (3.22) is satisfied.

Furthermore, for increasing the stability, $e(K_e)$ must be enlarged. Thus, K_e which results in maximum $e(K_e)$ can reach the highest stability. Also, the highest stability leads to the best signal correction performance. Hence, K_e which maximizes $e(K_e)$ should be chosen to achieve the lowest steady-state error.

3.3 Overall Control Block Diagram

The overall control block diagram of the proposed improved MPC for the *LC*-HAPF is illustrated in Fig. 3.4, where BPF and HPF stand for bandpass filter and high-pass filter, respectively, and their filter parameters are shown in Table 3.2. f_{cb} and B denote the center frequency and bandwidth of the BPF, respectively, and f_{ch} denotes the cutoff frequency of the HPF. In Fig. 3.3, v_{xf} stands for the bandpass-filtered fundamental component of v_x and the subscript 'y' denotes phase α, β and 0 in α-β-0 coordinate. According to the instantaneous p-q theory [1], the reference compensating current $i_{cx}{}^*$ for the *LC*-HAPF can be determined as below:

$$\begin{bmatrix} i_{c0}{}^* \\ i_{c\alpha}{}^* \\ i_{c\beta}{}^* \end{bmatrix}=\frac{1}{v_0\left(v_\alpha^2+v_\beta^2\right)}\begin{bmatrix} v_\alpha^2+v_\beta^2 & 0 & 0 \\ 0 & v_0v_\alpha & -v_0v_\beta \\ 0 & v_0v_\beta & v_0v_\alpha \end{bmatrix}\begin{bmatrix} p_0 \\ \tilde{p}_{\alpha\beta} \\ q_{\alpha\beta} \end{bmatrix} \tag{3.23}$$

where $i_{c\alpha}{}^*$, $i_{c\beta}{}^*$ and $i_{c0}{}^*$ are the $i_{cx}{}^*$ in α-β-0 coordinate, and v_α, v_β and v_0 are the v_x in α-β-0 coordinate. Moreover, p_0 is the zero-sequence instantaneous active power, $p_{\alpha\beta}$ is the instantaneous active power and $q_{\alpha\beta}$ is the instantaneous reactive power. Also, the $\tilde{p}_{\alpha\beta}$ is the negative-sequence instantaneous active power, and it can be obtained by high-pass-filtering the $p_{\alpha\beta}$. Also, $p_{\alpha\beta}$ and $q_{\alpha\beta}$ can be calculated in (3.24) based on [13].

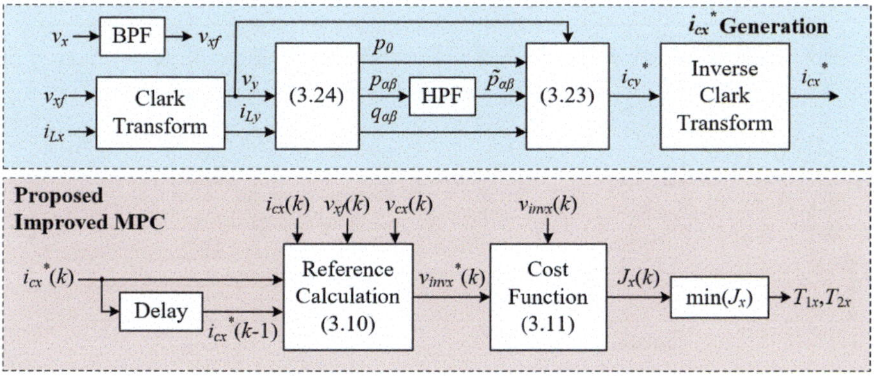

Fig. 3.4 Overall control block diagram of *LC*-HAPF with proposed improved MPC

$$\begin{bmatrix} p_0 \\ p_{\alpha\beta} \\ q_{\alpha\beta} \end{bmatrix} = \begin{bmatrix} v_0 & 0 & 0 \\ 0 & v_\alpha & v_\beta \\ 0 & -v_\beta & v_\alpha \end{bmatrix} \begin{bmatrix} i_{L0} \\ i_{L\alpha} \\ i_{L\beta} \end{bmatrix} \tag{3.24}$$

where $i_{L\alpha}$, $i_{L\beta}$ and i_{L0} are the i_{Lx} in α-β-0 coordinate. Finally, the reference compensating current i_{cx}^* can be produced by using (3.23, 3.24) and the inverse Clarke transformation. Then, the i_{cx}^* is input into the proposed improved MPC to generate the trigger signals for the switching devices of the VSI of the *LC*-HAPF.

3.4 Simulation Results

From (3.22) and the system parameters in Table 3.2, $K_e \geq$ -0.2 in order to ensure the system is stable. If K_e = -0.3 is applied to the proposed improved MPC for the *LC*-HAPF, the oscillation phenomenon appears and the compensation performance will be degraded (worst THDi_{sx} = 9.5%), as verified by Fig. 3.5a. Figure 3.5b presents the simulation results with K_e = -0.1, which shows better compensation performance (worst THDi_{sx} = 8.3%) without any oscillation phenomenon. Therefore, Fig. 3.5 verifies the effectiveness of the stable K_e range design (3.22) as discussed in Sect. 3.2.4.

To show the effectiveness of the proposed signal correction technique, Figs. 3.6 and 3.7 illustrates the *LC*-HAPF simulation results of the worst THDi_{sx} and settling time T_{st} versus K_e by using MATLAB/Simulink. The data is collected from 51 times simulation by using the improved MPC controlled *LC*-HAPF system with different K_e varies from -0.2 to 4.8 with a step size of 0.1, which satisfies the stability requirement of $K_e \geq$ -0.2. Figure 3.6 indicates that the worst THDi_{sx} increases significantly when K_e < 0.2, and the lowest worst THDi_{sx} appears at K_e = 0.7. Moreover, when K_e > 0.7, the worst THDi_{sx} also increases gradually. Hence, K_e = 0.7 is selected to acquire the smallest steady-state error. Figure 3.7 indicates that K_e

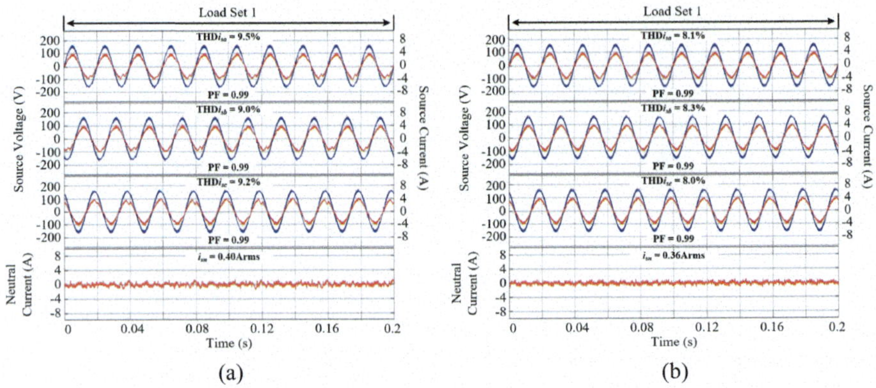

Fig. 3.5 Simulated results after *LC*-HAPF compensation under Load Set 1 with: **a** improved MPC with K_e = -0.3; **b** improved MPC with K_e = -0.1

does not affect the T_{st} much due to the direct optimal switching state selection of the MPC.

After that, simulation results of applying different controllers for the *LC*-HAPF are provided for comparison with the proposed improved MPC, and the system parameters are given in Table 3.2, where H is the hysteresis band of the HCC. K_p is the proportional gain, K_r is the resonant gain, and ω_c is the bandwidth of the resonant frequency of the QPRC. In addition, the DC-link voltage is supported by DC voltage source to simplify the verification. The simulation results of three-phase source voltage, source current and neutral current before *LC*-HAPF compensation under two load sets are show in Fig. 3.8, and the reference compensating current i_{cx}^* under two loads sets are shown in Fig. 3.9. The simulation results before and after the *LC*-HAPF compensation with different controllers are demonstrated in Figs. 3.10 and 3.11 and summarized in Table 3.3. Also, since two load sets are three-phase balanced and the equipment of the laboratory is limited, to ensure the consistence

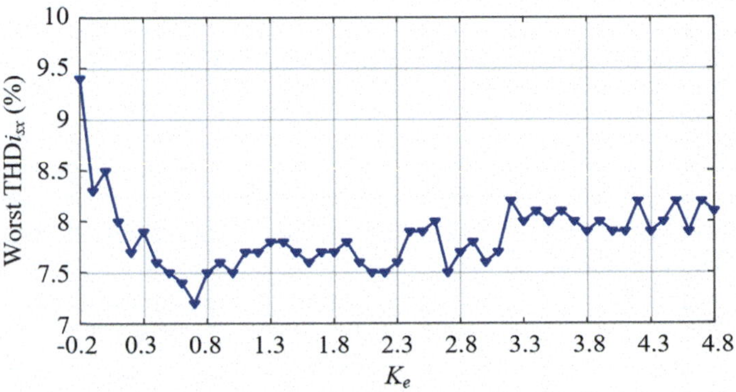

Fig. 3.6 The worst compensated THDi_{sx} of the *LC*-HAPF system with different K_e

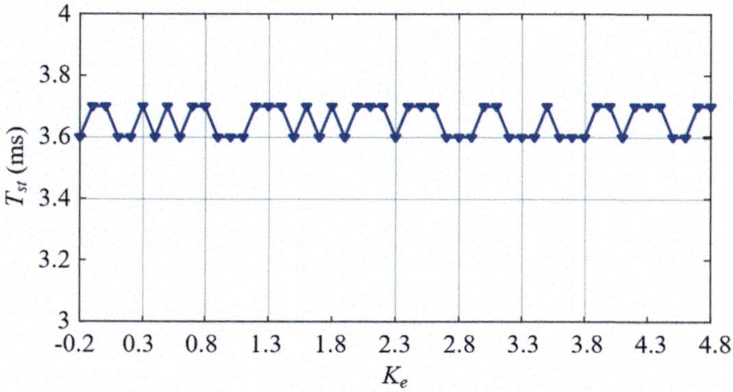

Fig. 3.7 The settling time T_{st} of the *LC*-HAPF system with different K_e

of the simulation and experimental results, only phase *a* waveforms are given in simulation results.

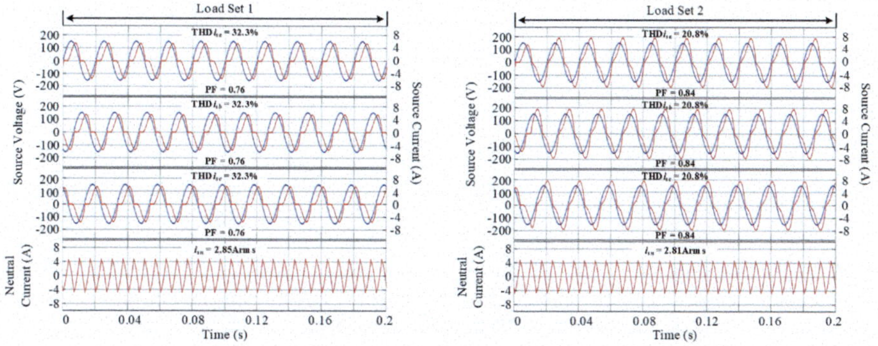

(a) (b)

Fig. 3.8 Simulated results of three-phase source voltage, source current and neutral current before *LC*-HAPF compensation under: **a** Load Set 1; **b** Load Set 2

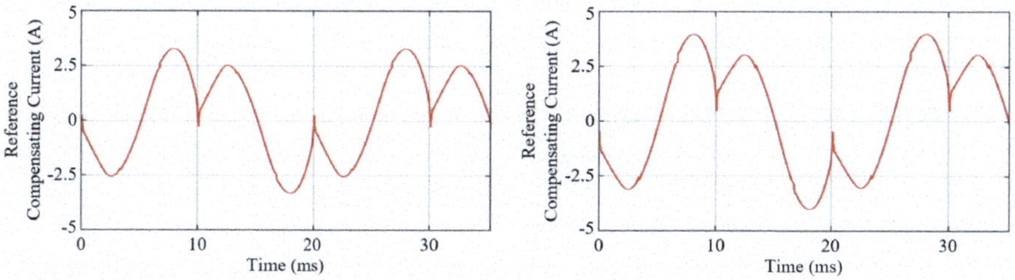

(a) (b)

Fig. 3.9 The simulated reference compensating current under: **a** Load Set 1; **b** Load Set 2

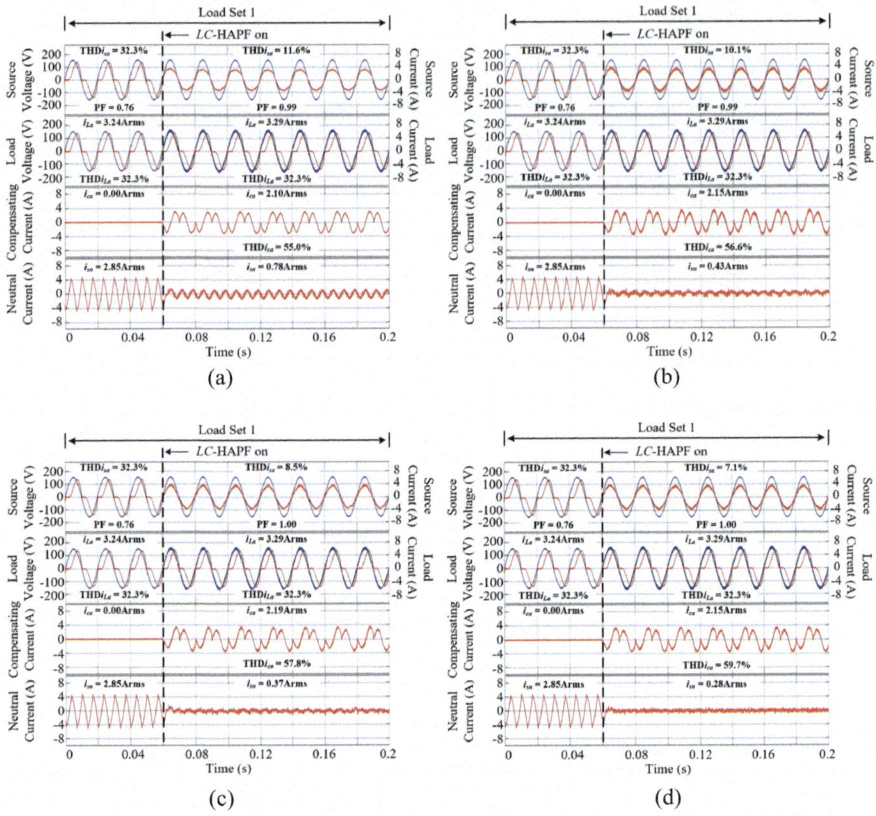

Fig. 3.10 Simulated results of phase a source voltage, source current, load voltage, load current, compensating current, and neutral current before and after *LC*-HAPF compensation under Load Set 1 with: **a** QPRC; **b** HCC; **c** conventional MPC; **d** improved MPC

(A) Quasi-Proportional-Resonant Controller: From Fig. 3.10a, 3.11a and Table 3.3, the QPRC has the worst current harmonics compensation performance and bad dynamic performance. The worst THDi_{sx} = 11.7% and 8.5% after compensation under Load Set 1 and Load Set 2, respectively. Besides, it has the highest compensated neutral current i_{sn} = 0.78Arms and 0.72Arms under Load Set 1 and Load Set 2, respectively.

(B) Hysteresis Current Controller: From Figs. 3.10b, 3.11b and Table 3.3, the HCC obtains a better harmonics compensation performance compared with the QPRC and good dynamic performance. The worst THDi_{sx} = 10.1% and 7.3% after compensation under Load Set 1 and Load Set 2, respectively. Besides, it has the second highest compensated neutral current i_{sn} = 0.43Arms and 0.52Arms under Load Set 1 and Load Set 2, respectively.

(C) Conventional Model Predictive Controller: From Figs. 3.10c, 3.11c and Table 3.3, the harmonics compensation performance of the conventional MPC is

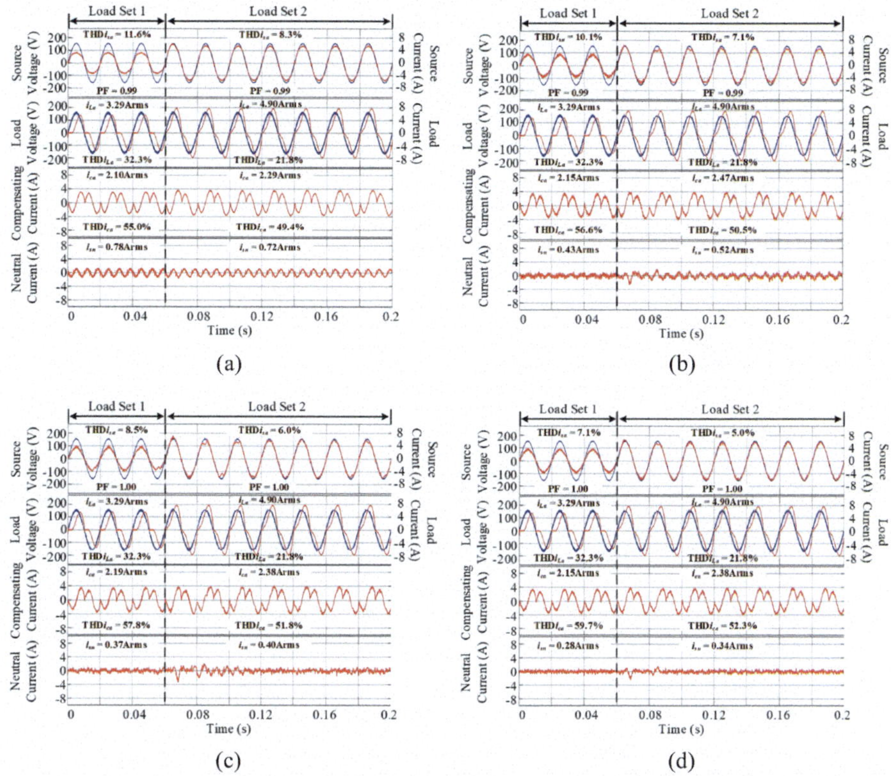

Fig. 3.11 Simulated results of phase a source voltage, compensated source current, load voltage, load current, compensating current, and neutral current under a step load change by *LC*-HAPF compensation with:**a** QPRC; **b** HCC; **c** conventional MPC; **d** improved MPC

better than the HCC, and its dynamic performance is similar to that of the HCC. The worst THDi_{sx} = 8.5% and 6.1% after compensation under Load Set 1 and Load Set 2, respectively. Besides, it has the second lowest i_{sn} = 0.37Arms and 0.40Arms under Load Set 1 and Load Set 2, respectively.

(D) Improved Model Predictive Controller: From Figs. 3.10d, 3.11d and Table 3.3, the improved MPC has the best current harmonics compensation performance and good dynamic performance. The worst THDi_{sx} = 7.2% and 5.0% after compensation under Load Set 1 and Load Set 2, respectively, which are about 1.3% and 1.1% lower than the conventional MPC, respectively. Besides, it has the lowest i_{sn} = 0.28Arms and 0.34Arms under Load Set 1 and Load Set 2, respectively. Hence, the proposed improved MPC can achieve small steady-steady error and fast dynamic response.

(E) Unbalanced Load Compensation: Under unbalanced load situation and within the *LC*-HAPF compensation range consideration, Fig. 3.12 and Table 3.4 also show the simulation results by the *LC*-HAPF compensation with the QPRC,

Table 3.3 Simulation results before and after *LC*-HAPF compensation

Load	Performance Index	Before Compensation	After Compensation			
			QPRC	HCC	Conventional MPC	Improved MPC
Load Set 1	Worst THDi_{sx} (%)	32.3	11.7	10.1	8.5	7.2
	PF	0.76	0.99	0.99	1.00	1.00
	Avg. f_{sw} (kHz)	/	10.0	4.1	4.2	4.4
	Total SW loss (W)	/	27.0	24.0	24.0	25.8
	i_{sn} (Arms)	2.85	0.78	0.43	0.37	0.28
	Q (var)	204	2	3	3	3
	i_{sx} (Arms)	3.25	2.46	2.49	2.52	2.53
Load Set 2	Worst THDi_{sx} (%)	20.8	8.5	7.3	6.1	5.0
	PF	0.84	0.99	0.99	1.00	1.00
	Avg. f_{sw} (kHz)	/	10.0	3.4	4.1	4.3
	Total SW loss (W)	/	22.2	18.0	18.6	19.8
	i_{sn} (Arms)	2.81	0.72	0.52	0.40	0.34
	Q (var)	266	6	10	10	10
	i_{sx} (Arms)	4.84	4.14	4.21	4.27	4.27

HCC, conventional MPC, and improved MPC, which clearly show that the improved MPC controlled *LC*-HAPF obtains the fast transient response and the best current harmonics, reactive power, unbalanced and neutral current compensation. In addition, the phasor diagrams of the unbalanced load before and after the *LC*-HAPF compensation are shown in Figs. 3.13 and 3.14, where v_{sx} denotes source voltage.

3.5 Experimental Results

To verify the simulation results, a 110V-5kVA three-phase four-wire center-split *LC*-HAPF experimental prototype is constructed in the laboratory as demonstrated in Fig. 2.11. A digital signal processor (DSP) TMS320F2812 is used as the control system with sampling frequency $f_s = 10$ kHz. For the VSI part, Mitsubishi IGBT intelligent power modules PM300DSA060 is used as the switching devices. Lastly, the experimental results of three-phase source voltage, source current and neutral current before *LC*-HAPF compensation under two load sets are show in Fig. 3.15, and the reference compensating current i_{cx}^* under two loads sets are shown in Fig. 3.16.

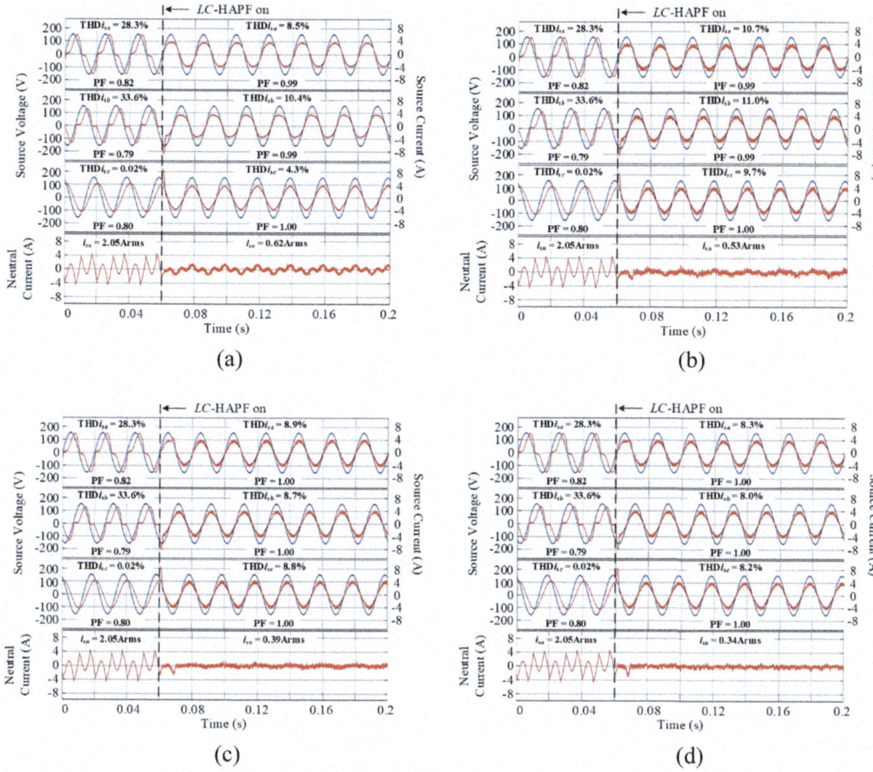

Fig. 3.12 Simulated results of three-phase source voltage, source current, and neutral current before and after *LC*-HAPF compensation under unbalanced load situation with: **a** QPRC; **b** HCC; **c** conventional MPC; **d** improved MPC

The experimental results before and after the compensation by the *LC*-HAPF with different controllers are summarized in Fig. 3.17, 3.18, 3.19, 3.20 and Table 3.3. Given that two load sets are three-phase balanced, only phase *a* waveforms are given in experimental results due to the limited equipment in the laboratory.

Additionally, based on the IEEE standard 519–2014 [14], the acceptable Total Demand Distortion (TDD) $\leq$ 15% with I_{SC}/I_L which is in 100 < 1000 scale (a small experimental rating 110V-5kVA *LC*-HAPF prototype). Moreover, the nominal rate current is assumed to be equal to the fundamental load current at the worst-case analysis, and this results in THD = TDD $\leq$ 15%. Hence, in this chapter, the performance of *LC*-HAPF current harmonics compensation is evaluated by setting an acceptable THD $\leq$ 15%.

(A) Quasi-Proportional-Resonant Controller: From Figs. 3.17a, 3.18, 3.19, 3.20a and Table 3.5, the QPRC has the worst current harmonics compensation performance and bad dynamic performance. The worst THDi_{sx} = 12.3% and 9.5% after compensation under Load Set 1 and Load Set 2, respectively. Besides, it has the

Table 3.4 Simulation results before and after *LC*-HAPF compensation under unbalanced load

Performance Index	Phase	Before Compensation	After Compensation			
			QPRC	HCC	Conventional MPC	Improved MPC
THDi_{sx} (%)	*a*	28.3	8.5	10.7	8.9	8.3
	b	33.6	10.4	11.0	8.7	8.0
	c	0.02	4.3	9.7	8.8	8.2
PF	*a*	0.82	0.99	0.99	1.00	1.00
	b	0.79	0.99	0.99	1.00	1.00
	c	0.80	1.00	1.00	1.00	1.00
i_{sn} (A_{rms})	/	2.05	0.62	0.53	0.39	0.34
Q (var)	*a*	203	23	17	10	-4
	b	180	-14	-10	-8	4
	c	205	-9	-7	-8	-4
i_{sx} (A_{rms})	*a*	3.67	2.73	2.71	2.70	2.69
	b	3.13	2.66	2.70	2.66	2.70
	c	3.08	2.61	2.63	2.72	2.70

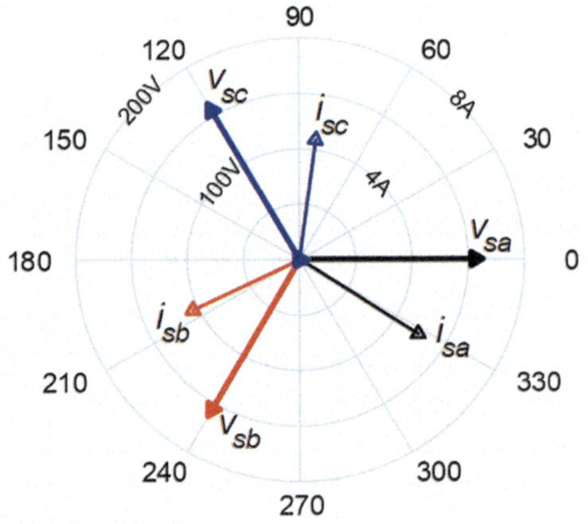

Fig. 3.13 Phasor diagram before *LC*-HAPF compensation under unbalanced load condition

highest compensated neutral current i_{sn} = 1.2Arms and 1.0Arms under Load Set 1 and Load Set 2, respectively.

(B) Hysteresis Current Controller: From Figs. 3.17b, 3.18, 3.19, 3.20b and Table 3.5, the HCC obtains a better harmonics compensation performance compared with the QPRC and good dynamic performance. The worst THDi_{sx} = 10.6% and 8.3% after compensation under Load Set 1 and Load Set 2, respectively. Besides, it has the second highest compensated neutral current i_{sn} = 1.0Arms and 0.7Arms under Load Set 1 and Load Set 2, respectively.

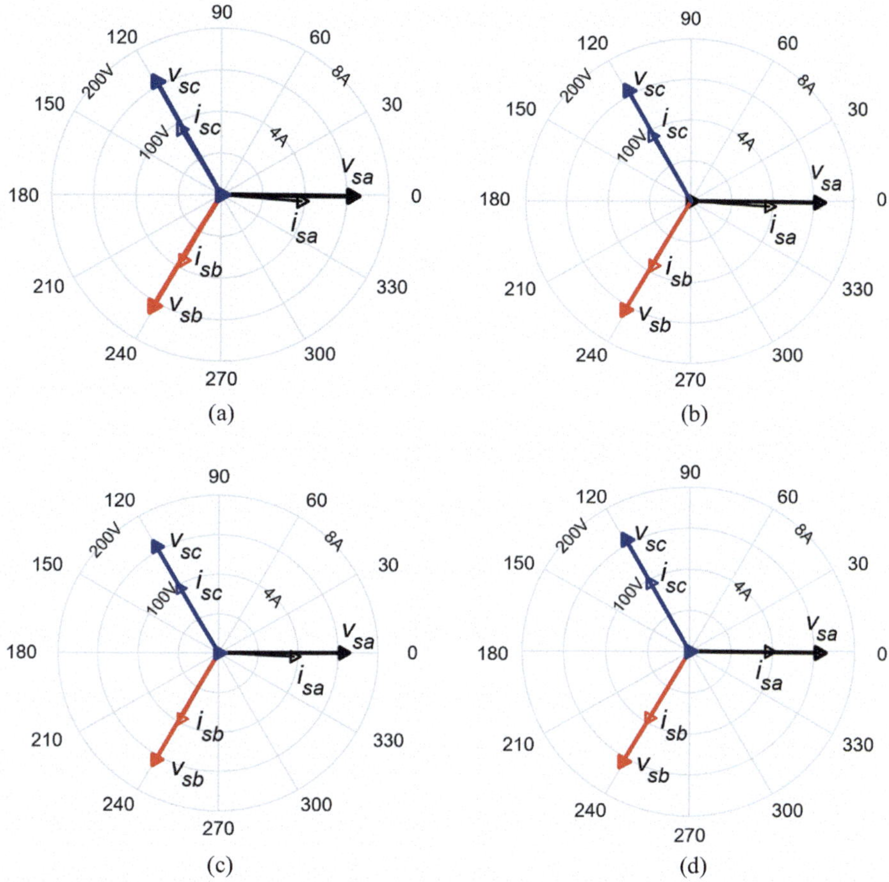

Fig. 3.14 Phasor diagram after *LC*-HAPF compensation under unbalanced load condition with: **a** QPRC; **b** HCC; **c** conventional MPC; **d** improved MPC

(C) Conventional Model Predictive Controller: From Figs. 3.17c, 3.18, 3.19, 3.20c and Table 3.5, the harmonics compensation performance of the conventional MPC is better than the HCC, and its dynamic performance is similar to that of the HCC. The worst THDi_{sx} = 8.9% and 6.8% after compensation under Load Set 1 and Load Set 2, respectively. Besides, it has the second lowest i_{sn} = 0.9Arms and 0.6Arms under Load Set 1 and Load Set 2, respectively.

(D) Improved Model Predictive Controller: From Figs. 3.17d, 3.18, 3.19, 3.20d and Table 3.5, the improved MPC has the best current harmonics compensation performance and good dynamic performance. The worst THDi_{sx} = 7.6% and 5.9% after compensation under Load Set 1 and Load Set 2, respectively, which are about 1.3% and 0.9% lower than the conventional MPC, respectively. Besides, it has the lowest i_{sn} = 0.8Arms and 0.5Arms under Load Set 1 and Load Set 2, respectively.

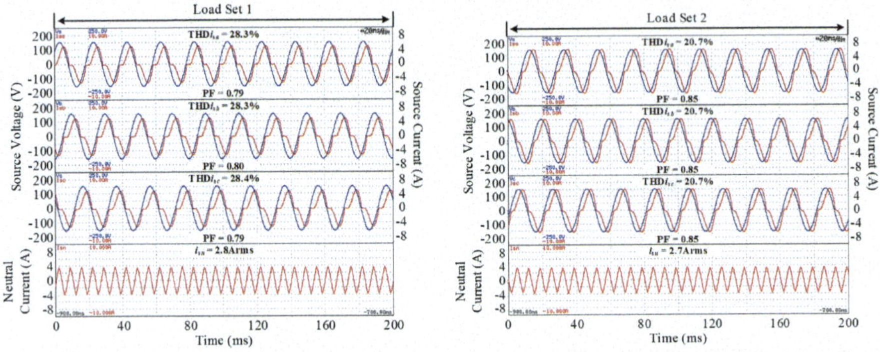

(a) (b)

Fig. 3.15 Experimental results of three-phase source voltage, source current and neutral current before *LC*-HAPF compensation under: **a** Load Set 1; **b** Load Set 2

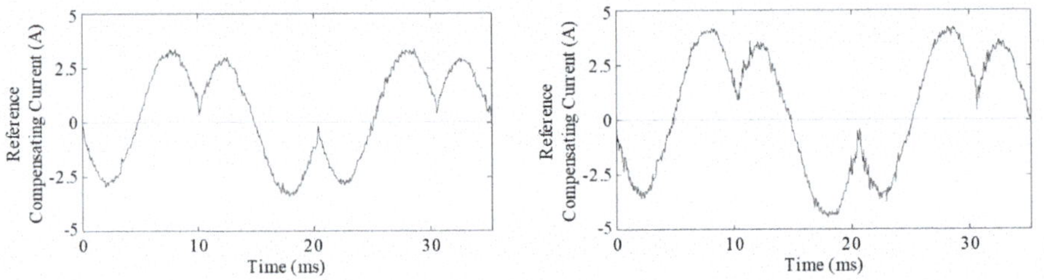

(a) (b)

Fig. 3.16 The reference compensating current during experiment under: **a** Load Set 1; **b** Load Set 2

Hence, the proposed improved MPC can achieve small steady-steady error and fast dynamic response.

(E) Comparison between Simulation and Experiment: The experimental results as shown in Figs. 3.17, 3.18, 3.19, 3.20 and Table 3.5 are consistent with the simulation results as shown in Figs. 3.10, 3.11 and Table 3.4, which demonstrates the validity and effectiveness of the proposed model predictive controller for the three-phase four-wire *LC*-HAPF. Moreover, the differences between them are caused by the sensor and control circuit, including the interference from the surrounding EMF, the resolution of the transducers, etc.

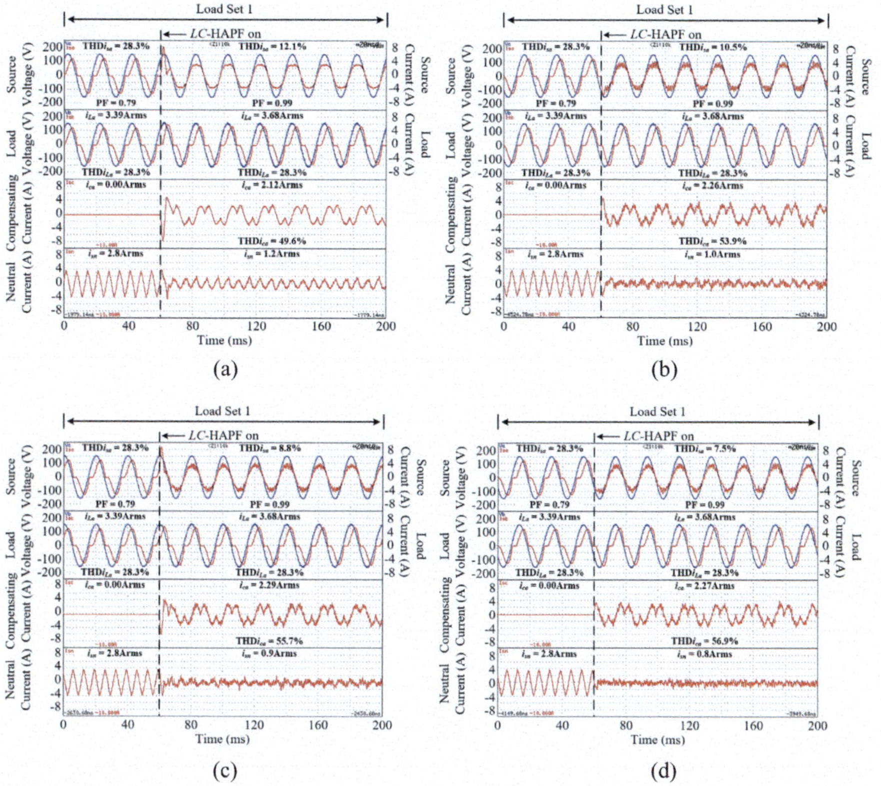

Fig. 3.17 Experimental results of phase a source voltage, source current, load voltage, load current, compensating current, and neutral current before and after *LC*-HAPF compensation under Load Set 1 with: **a** QPRC; **b** HCC; **c** conventional MPC; **d** improved MPC

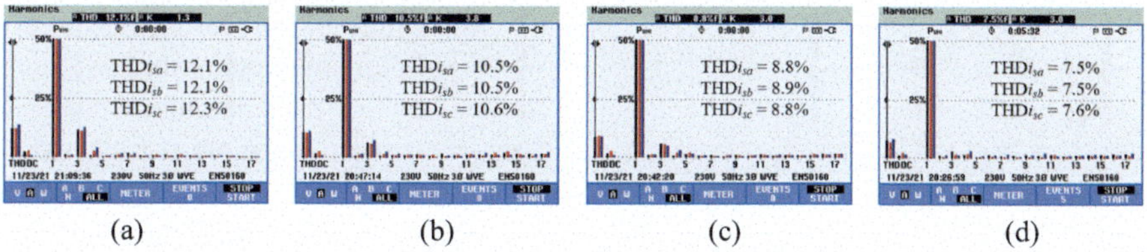

Fig. 3.18 Experimental results of three-phase compensated source current harmonic spectrums under Load Set 1 by *LC*-HAPF compensation with: **a** QPRC; **b** HCC; **c** conventional MPC; **d** improved MPC

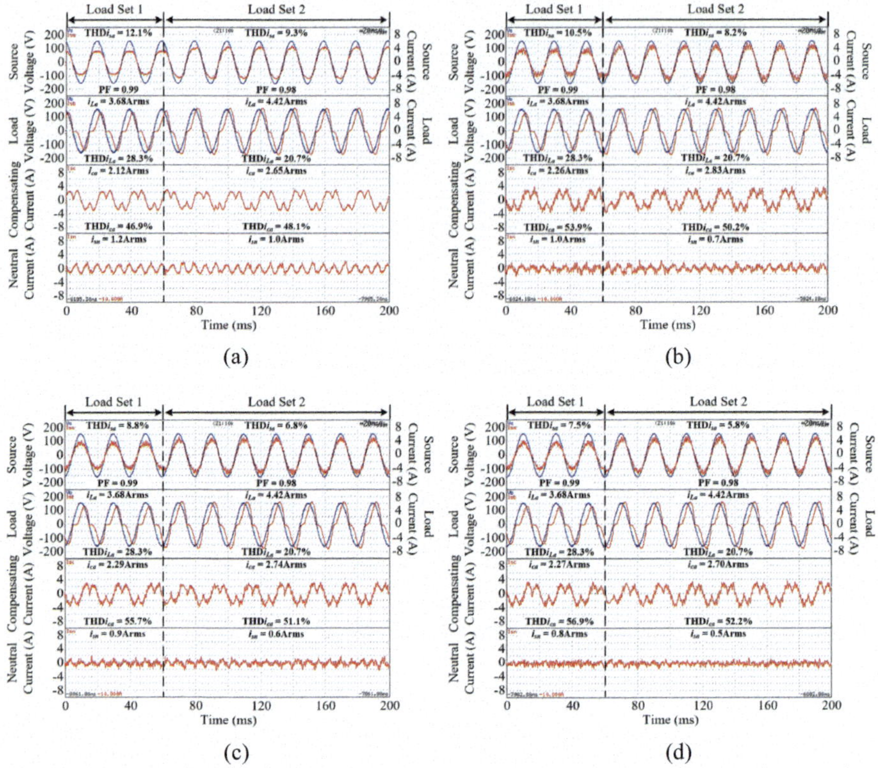

Fig. 3.19 Experimental results of phase a source voltage, compensated source current, load voltage, load current, compensating current, and neutral current under a step load change by *LC*-HAPF compensation with: **a** QPRC; **b** HCC; **c** conventional MPC; **d** improved MPC

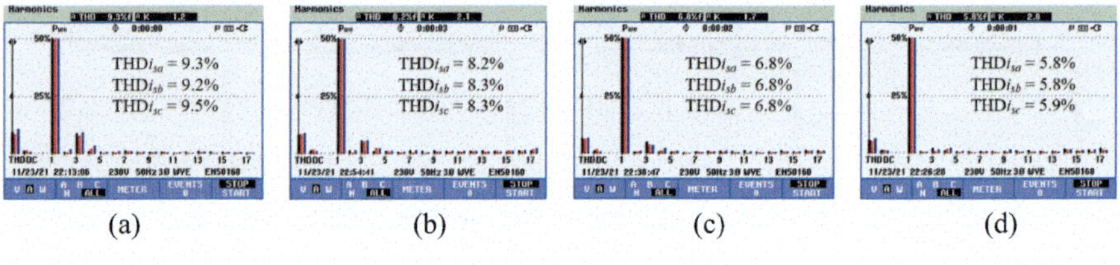

Fig. 3.20 Experimental results of three-phase compensated source current harmonic spectrums under Load Set 2 by *LC*-HAPF compensation with: **a** QPRC; **b** HCC; **c** conventional MPC; **d** improved MPC

Table 3.5 Experimental results before and after *LC*-HAPF compensation

Load	Performance Index	Before Compensation	After Compensation			
			QPRC	HCC	Conventional MPC	Improved MPC
Load Set 1	Worst THDi_{sx} (%)	28.4	12.3	10.6	8.9	7.6
	PF	0.79	0.99	0.99	0.99	0.99
	Avg. f_{sw} (kHz)	/	10.0	3.6	3.8	3.9
	Total SW loss (W)	/	30.0	27.0	27.6	29.4
	i_{sn} (A_{rms})	2.8	1.2	1.0	0.9	0.8
	Q (var)	200	0	10	10	10
	i_{sx} (A_{rms})	3.6	3.2	3.3	3.4	3.4
Load Set 2	Worst THDi_{sx} (%)	20.7	9.5	8.3	6.8	5.9
	PF	0.85	0.99	0.99	0.99	0.99
	Avg. f_{sw} (kHz)	/	10.0	3.1	3.4	3.6
	Total SW loss (W)	/	24.6	21.0	22.2	23.4
	i_{sn} (A_{rms})	2.7	1.0	0.7	0.6	0.5
	Q (var)	260	10	20	20	20
	i_{sx} (A_{rms})	5.1	4.1	4.3	4.4	4.4

3.6 Summary

In this chapter, an improved MPC with signal correction technique for the *LC*-HAPF is proposed. It can provide fast dynamic response and low steady-state error. The system predictive model of the *LC*-HAPF for implementing the improved MPC is deduced. Moreover, the stability and the parameter design of the improved MPC for the *LC*-HAPF are proposed and studied with a detailed design guideline. In addition, the effectiveness of the proposed improved MPC for the *LC*-HAPF is verified by simulation and experimental results in comparison with the QPRC, HCC and the conventional MPC, which shows superior dynamic response and steady-state performance.

References

1. H. Akagi, S. Ogasawara and H. Kim, The theory of instantaneous power in three-phase four-wire systems: a comprehensive approach. *1999 IEEE Industry Applications Conference*, **1**, (1999), pp. 431–439
2. S. Ebrahimpanah, Q. Chen, L. Zhang, "Model predictive current control with duty cycle optimization for two-level three-phase grid-tied inverter with output LCL filter based on forward Euler approximation," *2017 International Conf. on Industrial Informatics—Computing Technology, Intelligent Technology, Industrial Information Integration*, (2017) pp. 155–158
3. M.B. Waheed, A.R. Bhatti, M. Amjad, Y. Saleem, S.A. Niazi, S. Khokhar, A simplified model predictive control of four-leg two-level inverter. Electr. Power Compon. Syst. **47**(14–15), 1287–1302 (2019)
4. R. Yang, J. Luo, J. Zhao and Z. Xu, "Model predictive control strategy based on prediction error feedforward for superheating process," *2015 10th Asian Control Conference*, (2015), pp. 1–6, 2015
5. J. Yoo, H.-S. Kim and S.-K. Sul, Design of frequency-adaptive flux observer in PMSM drives robust to discretization error. *IEEE Trans. on Ind. Electron.*, (May 2021)
6. G. Du, Z. Liu, F. Du, J. Li, Performance improvement of model predictive control using control error compensation for power electronic converters based on the Lyapunov function. J. Power Electron. **17**(4), 983–990 (2017)
7. R.C. Dorf, R.H. Bishop, *Modern control systems*, 13th ed., New York City: Pearson, (2016), pp. 371–379
8. R.C. Dorf, R.H. Bishop, *Modern control systems*, 13th ed., New York City: Pearson, (2016), pp. 602–613
9. J.E. Slotine, W. Li, *Applied nonlinear control*, 1st ed., New York City: Pearson, (1990), pp. 47–52
10. S. Kwak, S.-J. Yoo, J. Park, Finite control set predictive control based on Lyapunov function for three-phase voltage source inverters. IET Power Electron. **7**(11), 2726–2732 (2014)
11. Y. Zhang, G. Du, J. Li, Y. Lei, Hybrid control strategy of MPC and DBC to achieve a fixed frequency and superior robustness. Energies **13**(5), 1176 (2020)
12. C.-S. Lam, X.-X. Cui, W.-H. Choi, M.-C. Wong, Y.-D. Han, Minimum inverter capacity design for LC-hybrid active power filters in three-phase four-wire distribution systems. IET Power Electron. **5**(7), 956–968 (2012)
13. H. Akagi, Y. Kanazawa, A. Nabae, "Instantaneous reactive power compensators comprising switching devices without energy storage components," *IEEE Trans. on Ind. Appl.*, **IA-20**(3), (1984) pp. 625–630
14. IEEE Standard 519–2014: "IEEE recommended practices and requirements for harmonic control in electrical power systems", (2014)

Chapter 4
Vector Proportional Integral Control for *LC*-HAPF Under Minimum DC-Link Voltage

4.1 Introduction

In this chapter, the vector proportional integral control (VPIC) for *LC*-HAPF with ideally zero steady-state error, optimal transient response and good robustness is proposed. First, the transfer function of the proposed VPIC is introduced, and its parameter design for obtaining optimal transient response is discussed. Secondly, a detailed robustness analysis of the VPIC controller including the system parameter variation, distorted grid voltage and grid frequency variation conditions is proposed. Thirdly, the digital implementation method of the VPIC and an overall control block diagram for the *LC*-HAPF with the proposed VPIC are presented. Moreover, a comparative study and analysis between the VPIC and the other representative current controllers under the minimum DC-link voltage is discussed, in which the VPIC obtains the best performances under the minimum DC-link voltage operation. Simulation and experimental results are provided to verify the advantages and effectiveness of the proposed VPIC for a three-phase four-wire *LC*-HAPF in comparison with the other representative current controllers under different DC-link voltage conditions. Finally, the robustness under different grid conditions and unbalanced compensation capability of the proposed VPIC controlled *LC*-HAPF is also verified by experimental results.

The contribution of this chapter is summarized as follows:

(1) Propose a VPIC and its parameter design for the *LC*-HAPF with zero steady-state error, fast transient response, and good robustness.
(2) Propose a detailed robustness analysis of the VPIC including the system parameter variation, distorted grid voltage and grid frequency variation conditions.
(3) Propose an overall control block diagram for the *LC*-HAPF with the proposed VPIC and the digital implementation method of the VPIC.
(4) Propose a comparative study and analysis of the VPIC with other representative current controllers under minimum DC-link voltage operation.

C.-S. Lam et al., *Advanced Controllers Design for Hybrid Active Power Filters*,
https://doi.org/10.1007/978-981-97-6847-9_4

(5) Verify the advantages and effectiveness of the proposed VPIC for the *LC*-HAPF by simulation and experimental results in comparison with the QPRC, PMRC, and SOSMC. The robustness of the proposed VPIC controlled *LC*-HAPF under the system parameter variation, distorted grid voltage and grid frequency variation conditions is also validated.

4.2 System Modeling and Minimum DC-Link Voltage Design of *LC*-HAPF

4.2.1 *System Modeling of* LC-*HAPF*

From the circuit configuration of the three-phase four-wire *LC*-HAPF as shown in Fig. 1.4, the VPIC design based on the *LC*-HAPF system model as shown in below.

$$L_c \frac{di_{cx}}{dt} = v_x - v_{cx} - \frac{V_{dc}}{2} u_x. \tag{4.1}$$

$$C_c \frac{dv_{cx}}{dt} = i_{cx} \tag{4.2}$$

From (4.1) and (4.2), the *s*-domain block diagram of the current control loop of the *LC*-HAPF can be built as shown in Fig. 4.1. $C_{yx}(s)$ represents the different current controllers, $P_F(s)$ and $P_d(s)$ are the model of the *LC* filter and the time-delay of the uniformly sampled PWM with double update [1], which can be expressed as:

$$P_F(s) = \frac{C_c s}{L_c C_c s^2 + 1} \tag{4.3}$$

$$P_d(s) = e^{-2T_s s} \tag{4.4}$$

where s is the Laplace operator, T_s is the sampling period.

From Fig. 4.1, the open-loop transfer function $L_x(s)$ is derived as:

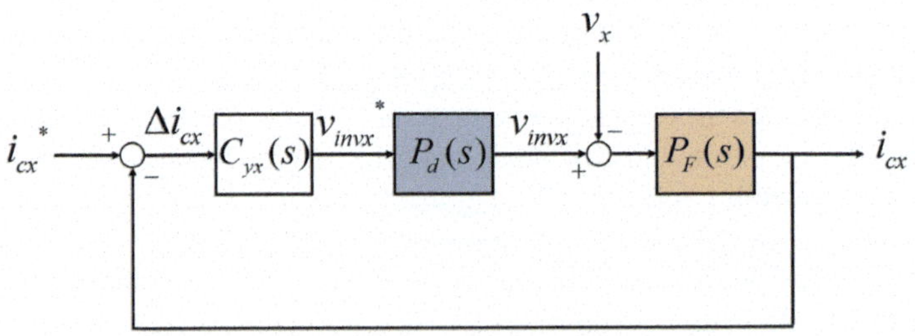

Fig. 4.1 *s*-domain block diagram of current control loop of the *LC*-HAPF

$$L_x(s) = C_{yx}(s)P_d(s)P_F(s) \tag{4.5}$$

Moreover, the output current can be derived as:

$$i_{cx}(s) = T_x(s)i_{cx}{}^*(s) - Y_x(s)v_x(s) \tag{4.6}$$

where $T_x(s)$ and $Y_x(s)$ are given as follows:

$$T_x(s) = \frac{L_x(s)}{1 + L_x(s)} \tag{4.7}$$

$$Y_x(s) = \frac{P_F(s)}{1 + L_x(s)} \tag{4.8}$$

where $T_x(s)$ is the closed-loop transfer function, and $Y_x(s)$ is the equivalent output impedance [2].

4.2.2 *Minimum DC-Link Voltage Design of* LC*-HAPF*

To optimize the *LC*-HAPF performances, the minimum DC-link voltage design can be utilized for switching loss and switching ripple reduction [3–5]. Assume that the voltages of upper and lower DC-link capacitors are approximately balanced, the steps of minimum DC-link voltage calculation are given in the following.

Firstly, the minimum DC-link voltage for compensating fundamental reactive power is derived by:

$$V_{dcf} = \left| v_x(j\omega_1) - \frac{i_{cx}{}^*(j\omega_1)}{P_F(j\omega_1)} \right| \tag{4.9}$$

where ω_1 is the fundamental angler frequency, V_{dcf} is the DC-link voltage requirement for compensating reactive power at the fundamental frequency.

Then, the minimum DC-link voltage for compensating each *n*-th order current harmonics can be calculated as:

$$V_{dcn} = \left| v_x(j\omega_n) - \frac{i_{cx}{}^*(j\omega_n)}{P_F(j\omega_n)} \right| \tag{4.10}$$

where ω_n is *n*th order harmonic angular frequency, V_{dcn} is the DC-link voltage requirement for compensating each *n*th order harmonic current.

Finally, the minimum DC-link voltage for compensating both fundamental reactive power and the harmonic current can be calculated by:

$$V_{dcm}=\sqrt{\left|V_{dcf}\right|^2+\sum_{n=2}^{h}\left|V_{dcn}\right|^2} \tag{4.11}$$

where h is the highest harmonic order taken consideration.

4.3 Vector Proportional Integral Controller Design of *LC*-HAPF

4.3.1 Derivation of VPIC

The primary goal of the controller design is to track the reference compensating current $i_{cx}{}^*$ with zero steady-state error. From (4.6) to (4.8), the steady-state tracking error Δi_{cx} at each harmonic frequency ω_n can be expressed as:

$$\begin{aligned}\Delta i_{cx}(j\omega_n) &= i_{cx}{}^*(j\omega_n) - i_{cx}(j\omega_n)\\ &= \left[1 - T_x(j\omega_n)\right]i_{cx}{}^*(j\omega_n) - Y_x(j\omega_n)v_x(j\omega_n)\end{aligned} \tag{4.12}$$

where the subscript n represents each harmonic order. To guarantee the steady-state error $\Delta i_{cx}(j\omega_n) = 0$ under arbitrary value of $i_{cx}{}^*(j\omega_n)$ and $v_x(j\omega_n)$, the following condition must be satisfied:

$$\begin{cases} T_x(j\omega_n) = 1\angle 0^{\circ} \\ Y_x(j\omega_n) = 0 \end{cases} \tag{4.13}$$

From (4.7) and (4.8), to satisfy (4.13), the condition is derived as:

$$L_x(j\omega_n) = \infty \tag{4.14}$$

To meet the condition (4.14), the solution of the open-loop transfer function $L_x(s)$ is given as follows:

$$L_x(s) = K_x L_{nx}(s) = K_x \sum_{n=1}^{h} \frac{s}{s^2 + \omega_n^2} \tag{4.15}$$

where L_{nx} is the open-loop transfer function for nth harmonic, K_x is the control gain parameter, which will be designed and discussed in the next subsection. The identical gain of the proposed VPIC aims to minimize the design complexity. After getting the open-loop transfer function $L_x(s)$, with the help of (4.5) and (4.15), the VPIC at the ideal case can be derived as below:

$$C_{VPIx}^{ideal}(s) = L_x(s)P_d^{-1}(s)P_F^{-1}(s) \tag{4.16}$$

Unfortunately, the inverse of time-delay P_d^{-1} (s) is non-causal, which cannot be implemented in practical case [6]. Thus, a lead compensator is used to substitute P_d^{-1} (s) as an approximation [7], which is derived as:

$$C_{lnx}(s) = \frac{\cos(2T_s\omega_n)s - \omega_h \sin(2T_s\omega_n)}{s} \tag{4.17}$$

Therefore, the proposed VPIC in practical case is obtained as:

$$C_{VPIx}(s) = K_x \sum_{n=1}^{h} L_{nx}(s)C_{lnx}(s)P_F^{-1}(s) \tag{4.18}$$

From (4.18), it is clearly shown that the transfer function of the VPIC for the *LC*-HAPF can be easily obtained by substituting the transfer functions (4.3), (4.15) and (4.17) into it. Although (4.15) and (4.17) are similar to the conventional multi-resonant controller, the VPIC in (4.18) is newly proposed. This is because the VPIC does not contain a proportional term as the conventional multi-resonant controller does, and the system inverse part is a specific design for the *LC*-HAPF which has not been proposed in the literature. Applying the proposed transfer function (4.18) to APF, and the resulted controller has the same formulation as the conventional VPIC proposed for APF in [8]. Based on this fact, the proposed controller for *LC*-HAPF is also named as VPIC.

4.3.2 Optimal Parameter Design of VPIC

Next, we will study the parameter design of the VPIC for stability analysis and optimizing the *LC*-HAPF transient response speed. The *LC*-HAPF system transient performance can be investigated and reflected by the closed-loop poles of the *LC*-HAPF system, which is the solution of the system characteristic equation:

$$1 + L_x(s) = 0 \tag{4.19}$$

Figure 4.2 shows the root-locus of the VPIC controlled *LC*-HAPF system, where the system parameters are shown in Table 4.1 and $h = 9$ is chosen here as an example. The root-locus provides all possible locations of the closed-loop poles with $K_x = 0 \rightarrow$ on the *s*-plane. The system is stable if and only if all its closed-loop poles are located on the left-half-plane (LHP).

The transient response speed can be evaluated by the system decay rate, which is the negative real part of each closed-loop pole on the *s*-plane [8]. The larger decay rate means the faster transient response speed. Therefore, the control gain K_x value that

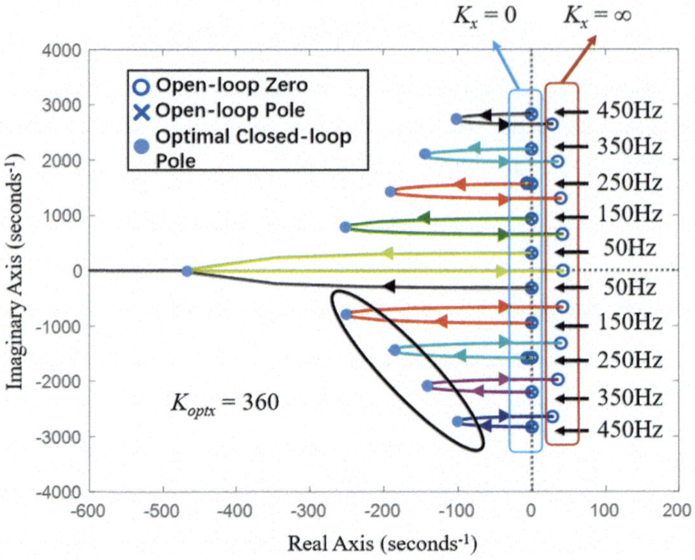

Fig. 4.2 Root locus of the closed-loop VPIC controlled *LC*-HAPF system

Table 4.1 System parameters for analysis, simulation and experiment

System Parameters	Physical Values
v_{gx}, f_g	110 V_{rms}, 50 Hz
L_s	1 mH
R_{NL}, L_{NL}, C_{NL} (Balanced nonlinear load)	45 Ω, 35 mH, 400 μF
$R_{NLa}, R_{NLb}, R_{NLc}, L_{NLx}, C_{NLx}$ (Unbalanced Nonlinear load)	30 Ω, 60 Ω, 40 Ω, 35 mH, 400 μF
L_c, C_c	8 mH, 50 μF
V_{dc}	50, 30 V
f_s	20 kHz
f_{sw}	10 kHz

can output the leftmost closed-loop pole on the *s*-plane will be the optimum control gain K_{optx} with maximum decay rate (optimal transient response speed). To find this optimum control gain K_{optx}, the root locus of the closed-loop VPIC controlled *LC*-HAPF system is illustrated in Fig. 4.2. Figure 4.2 shows that each root locus is a convex curve with a unique leftmost peak at the LHP of the *s*-plane, which is the closed-loop pole with the largest decay rate. It is worth noting that, the leftmost peak also locates at the farthest distance from the imaginary axis, which also represents the largest stability margin. Therefore, the optimum control gain K_{optx} can be obtained directly from the root-locus, which is $K_{optx} = 360$.

4.3.3 Robustness Analysis of VPIC

The performance of a control system can be reflected from its closed-loop frequency response [6]. And thus, the robustness of the *LC*-HAPF with VPIC can be verified from the bode diagrams of the closed-loop transfer function $T_x(s)$ and equivalent output impedance $Y_x(s)$.

Considering the worst-case frequency variation defined by IEEE Std. 446-1995 ($\pm 1\%$) [9], the closed-loop frequency responses $T_x(s)$ of the LC-HAPF with VPIC under 49.5, 50 and 50.5 Hz grid frequency are demonstrated in Fig. 4.3, in which they yield approximately the same frequency response.

When the filter parameter L_c or C_c is having $\pm 10\%$ variation, from Fig. 4.4, the closed-loop response $T_x(s)$ plots does not change much, which illustrates good robustness of the VPIC controlled *LC*-HAPF under filtering parameter variation.

The robustness of the *LC*-HAPF with VPIC under distorted grid voltage is reflected by the equivalent output impedance Y_x(s). From (4.8), the detailed formula of $Y_x(s)$ is derived as:

$$Y_x(s) = \frac{\frac{1}{L_c}\frac{s}{s^2+(1/L_cC_c)}}{1 + K_x \sum\limits_{n=1}^{h} \frac{s}{s^2+\omega_n^2}} \tag{4.20}$$

The bode diagram of the *LC*-HAPF output impedance without and with the VPIC is shown in Fig. 4.5, which shows that the *LC* resonance peak is eliminated by

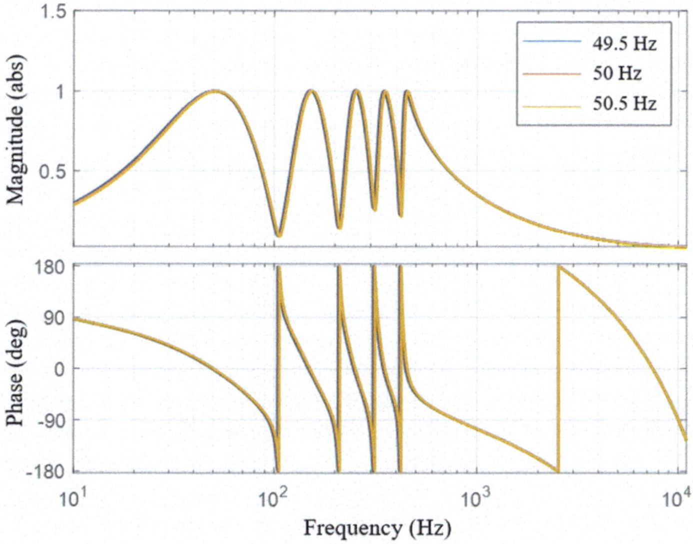

Fig. 4.3 Bode diagrams of the *LC*-HAPF closed-loop frequency responses with VPIC under different grid frequencies

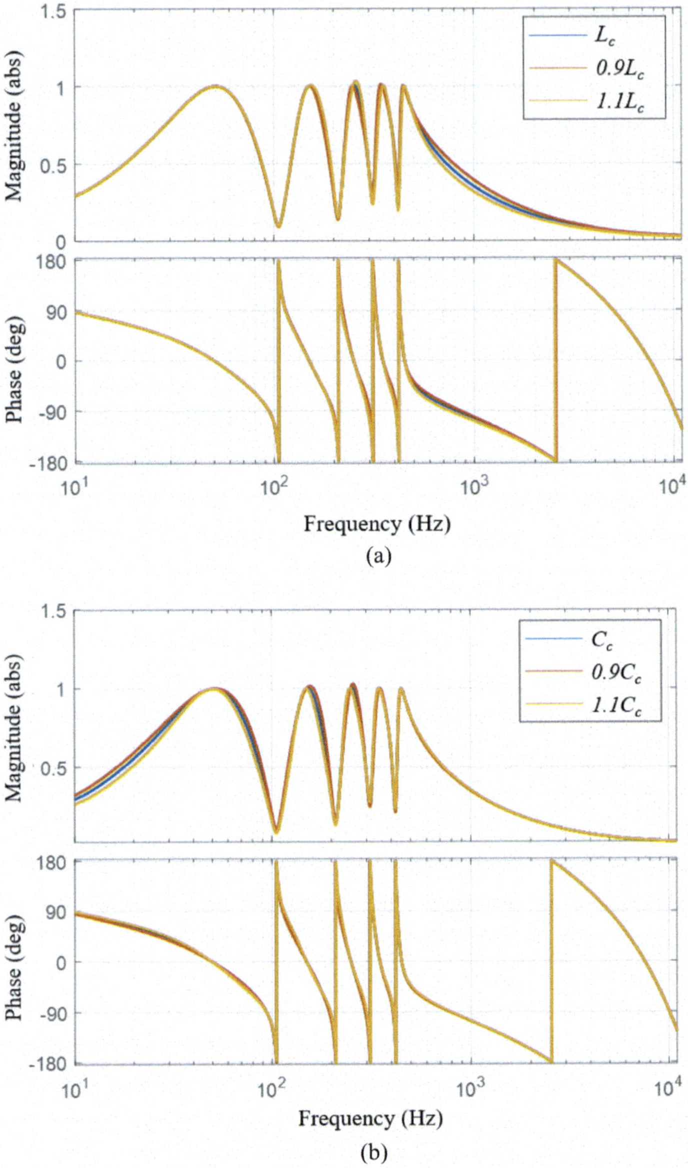

Fig. 4.4 Bode diagram of the *LC*-HAPF closed-loop response with the proposed VPIC under filtering parameter variation: **a** with ±10% L_c variation; **b** with ±10% C_c variation

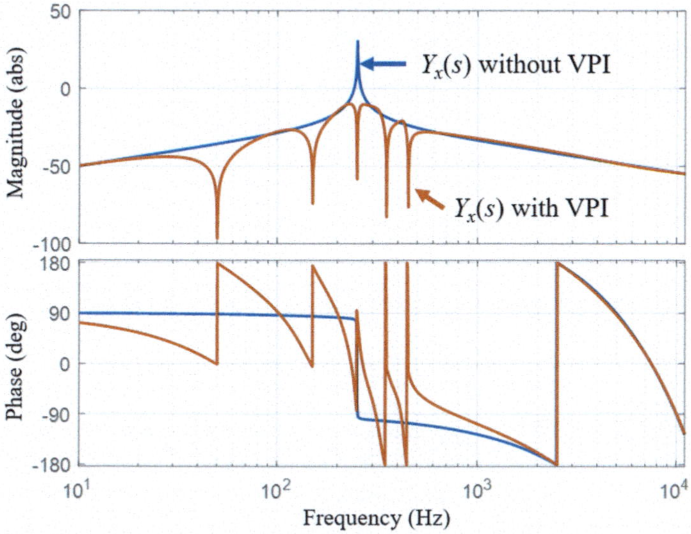

Fig. 4.5 Bode diagram of *LC*-HAPF output impedance without and with the VPIC

the proposed VPIC, which can damp the potential *LC* resonance introduced by the background harmonic in the grid voltage.

4.3.4 Digital Implementation of VPIC

The detailed formula of VPIC (4.18) is given in the following:

$$\begin{aligned} C_{VPIx}(s) &= K_x \sum_{n=1}^{h} L_{nx}(s) C_{Inx}(s) P_F^{-1}(s) \\ &= K_x \left[\begin{array}{l} \displaystyle\sum_{n=1}^{h} \left(\frac{\cos(2T_s\omega_n)s^2 - \omega_n \sin(2T_s\omega_n)s}{s^2 + \omega_n^2} \right) \\ \times \left(L_c + \dfrac{1}{C_c s^2} \right) \end{array} \right] \end{aligned} \tag{4.21}$$

For the sake of simplicity, the equation (4.21) can be decomposed as:

$$C_{VPIx}(s) = C_{VPIx}^1(s) C_{VPIx}^2(s) \tag{4.22}$$

where C_{VPIx}^1 and C_{VPIx}^2 can be represented by:

$$
\begin{aligned}
C^1_{VPIx}(s) &= K_x \sum_{n=1}^{h} \frac{k_{cn}s^2 + k_{sn}s}{s^2 + \omega_n^2} \\
&= K_x \sum_{n=1}^{h} \left[k_{cn} \frac{1}{1 + \frac{1}{s} \times \frac{1}{s} \times \omega_n^2} + k_{sn} \frac{\frac{1}{s}}{1 + \frac{1}{s} \times \frac{1}{s} \times \omega_n^2} \right]
\end{aligned} \tag{4.23}
$$

$$
C^2_{VPIx}(s) = L_c + \frac{1}{C_c s^2} \tag{4.24}
$$

where the auxiliary parameters k_{cn} and k_{sn} in (4.23) are defined as:

$$
k_{cn} = \cos(2T_s\omega_n),\ k_{sn} = -\omega_n \sin(2T_s\omega_n) \tag{4.25}
$$

The 1st part of the VPIC C^1_{VPIx} that contains multiple resonant terms are implemented by double integrators scheme. The forward integrator is discretized by using the forward Euler method, while the feedback integrator one is discretized by using the backward Euler method [10–13]. The derived transfer function C^1_{VPIx} in the z-domain can be expressed as:

$$
C^1_{VPIx}(z) = K_x \sum_{n=1}^{h} \left[\begin{array}{l} k_{cn} \dfrac{1}{1 + \frac{T_s z^{-1}}{1-z^{-1}} \times \frac{T_s}{1-z^{-1}} \times \omega_n^2} \\ + k_{sn} \dfrac{\frac{T_s z^{-1}}{1-z^{-1}}}{1 + \frac{T_s z^{-1}}{1-z^{-1}} \times \frac{T_s}{1-z^{-1}} \times \omega_n^2} \end{array} \right] \tag{4.26}
$$

where T_s is sampling frequency.

The 2nd part of the VPIC C^2_{VPIx} is implemented by using three integrators. All of the integrators contained in C^2_{VPIx} are discretized by using forward Euler method:

$$
C^2_{VPIx}(z) = L_c + \frac{1}{C_c} \left(\frac{T_s}{2} \frac{z^{-1}}{1 - z^{-1}} \right)^2 \tag{4.27}
$$

From (4.26) and (4.27), the block diagram of the digital implementation of the VPIC is shown in Fig. 4.6.

4.4 Overall Control Block Diagram

The overall control block diagram of the *LC*-HAPF with the proposed VPIC is provided in Fig. 4.7, which includes an instantaneous symmetrical component calculation block, an instantaneous power compensation control block, a DC-link voltage control block, and a final reference compensating current with VPIC block.

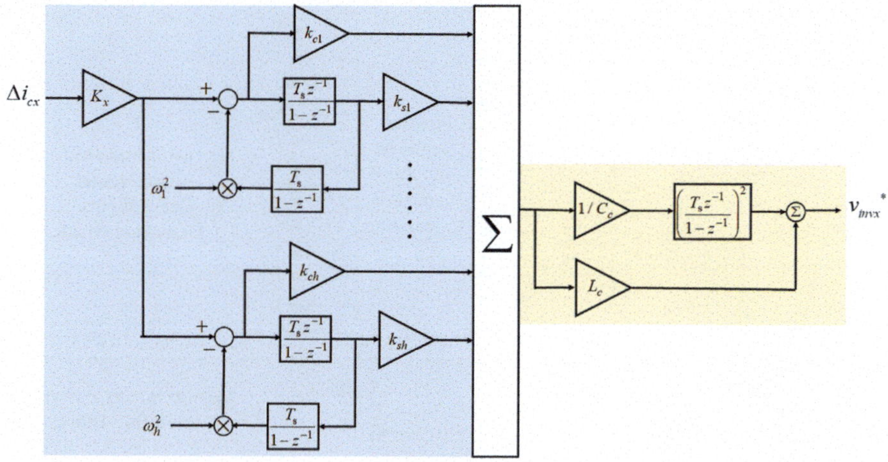

Fig. 4.6 Block diagram of the digital implementation of the proposed VPIC

The superscript "+", "−", and "0" denotes positive, negative and zero sequence component of the corresponding signal.

The instantaneous symmetrical component calculation block [14] is used to extract the positive, negative and zero sequence load current (i_{Lx}^+, i_{Lx}^-, i_{Lx}^0) and the fundamental positive sequence load voltage (v_x^+) signals in real-time from the measured load current and load voltage.

The instantaneous power compensation control block is used for unbalance, harmonic, and reactive power detection based on the instantaneous *p*-*q* theory [14]. The average value of positive sequence active power $p_{\alpha\beta}^+$ is filtered out by a low-pass filter, thus the remained signals containing the unbalance, harmonic, and reactive power are used for calculating the reference compensating currents in α-β-0 coordinate.

The DC-link voltage control block is used for providing a stable DC-link voltage. First, the sensed feedback DC-link voltage signal V_{dc} is compared with its reference value V_{dc}^*, and their difference will input to the proportional (P) or proportional integral (PI) controller to obtain the corresponding DC-link voltage control signals p_{dc} and q_{dc}. Then, these two control signals p_{dc} and q_{dc} will be added to the instantaneous compensating positive sequence active and reactive power references in the instantaneous power calculation block for calculating the reference compensating currents in α-β-0 coordinate. Then the final reference compensating currents (i_{xp}, i_{xq}) in the *a*-*b*-*c* coordinate can be obtained by the inverse matrix of Clarke transformation in α-β-0 coordinates. The detailed analysis and parameter design of the DC-link controller part has been reported in [15].

In the final reference compensating current with VPIC block, the compensating current error signal of each phase goes through the proposed VPIC and generates the reference inverter voltage v_{invx}. Finally, the voltage reference feeds into the PWM block and generates the trigger signals of the switching devices.

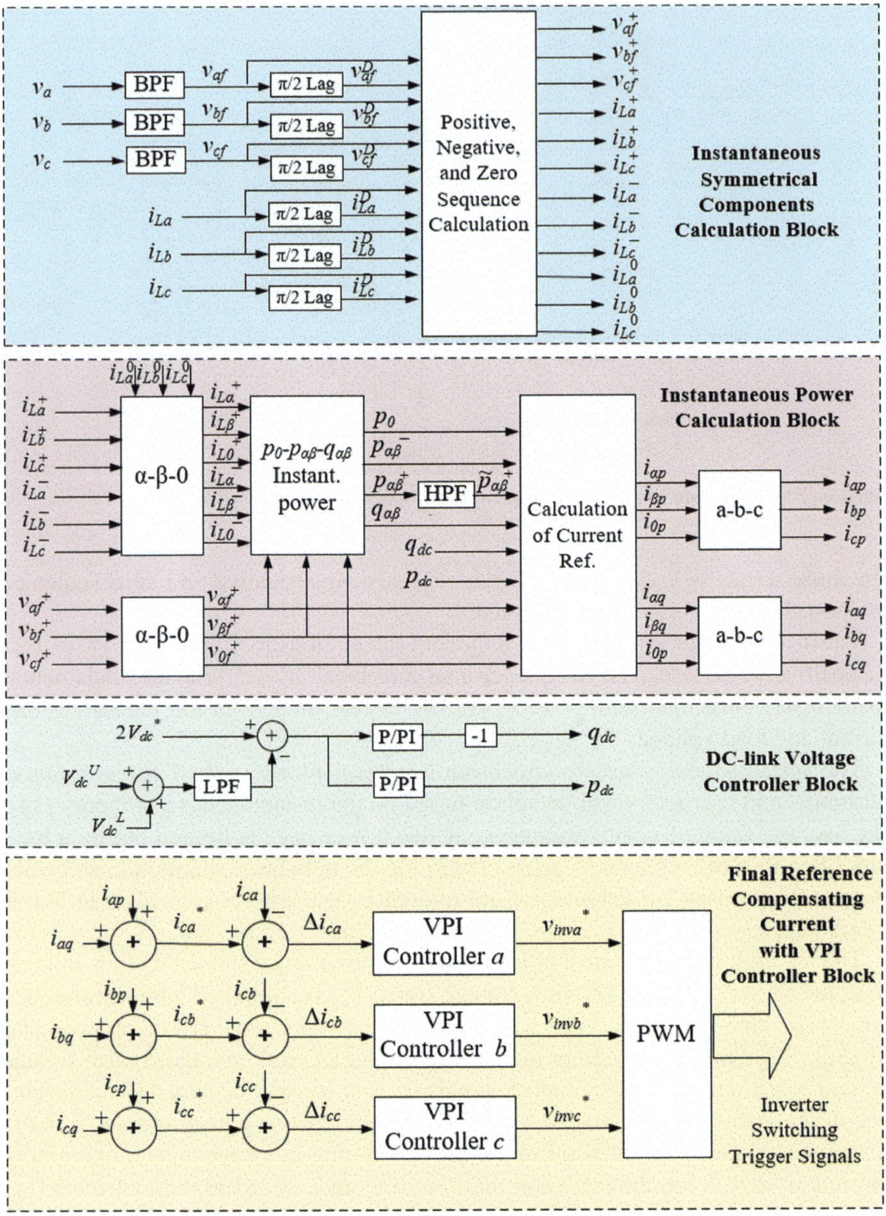

Fig. 4.7 Overall control block diagram of *LC*-HAPF with VPIC

4.5 Theoretical Comparative Study Among Different Controllers for *LC*-HAPF

4.5.1 Controllability of Current Controller with Minimum DC-Link Voltage

For the active inverter part of the *LC*-HAPF, the magnitude of the inverter voltage is always limited by its DC-link voltage:

$$|v_{invx}(t)| \leq v_{dc} \quad \forall t \geq 0. \tag{4.28}$$

If the inverter receives a voltage reference larger than its V_{dc}, the PWM will be overmodulated, and the output compensating current will be distorted. Thus, to obtain good compensation performance, the following condition must be satisfied [16]:

$$\left|v_{invx}{}^{*}(t)\right| \leq v_{dc} \quad \forall t \geq 0. \tag{4.29}$$

To calculate the magnitude of the left-hand side of (4.29), H_{∞} method in the frequency domain can be used here [16], and the condition (4.29) becomes:

$$\left\|v_{invx}{}^{*}(j\omega)\right\|_{\infty} \leq v_{dc}. \tag{4.30}$$

From Fig. 4.1, the voltage reference in *s*-domain can be derived as:

$$v_{invx}{}^{*}(s) = \frac{C_{yx}(s)}{1+L_x(s)} i_{cx}{}^{*}(s) - \frac{C_{yx}(s)P_F(s)}{1+L_x(s)} v_x(s). \tag{4.31}$$

Substituting (4.5), (4.7) into (4.31), the following equation can be obtained:

$$v_{invx}{}^{*}(s) = \frac{T_x(s)i_{cx}{}^{*}(s)}{P_d(s)P_F(s)} - \frac{T_x(s)}{P_d(s)} v_x(s). \tag{4.32}$$

Thus, the H_{∞} norm of $v_{invx}{}^{*}$ can be written as:

$$\begin{aligned}\left\|v_{invx}{}^{*}(j\omega)\right\|_{\infty} &= \left\|\frac{T_x(j\omega)i_{cx}{}^{*}(j\omega)}{P_d(j\omega)P_F(j\omega)} - \frac{T_x(j\omega)}{P_d(j\omega)} v_x(j\omega)\right\|_{\infty} \\ &\leq \left\|\frac{T_x(j\omega)}{P_d(j\omega)}\right\|_{\infty} \left\|\frac{i_{cx}{}^{*}(j\omega)}{P_F(j\omega)} - v_x(j\omega)\right\|_{\infty}.\end{aligned} \tag{4.33}$$

Considering $\|P_d(j\omega)\|_{\infty} = 1$, and by applying function norm inequality [17], (4.33) becomes:

$$\begin{aligned}\left\|v_{invx}{}^*(j\omega)\right\|_\infty &\le \|T_x(j\omega)\|_\infty \left\|\frac{i_{cx}{}^*(j\omega)}{P_F(j\omega)} - v_x(j\omega)\right\|_\infty \\ &\le \|T_x(j\omega)\|_\infty \left\|\frac{i_{cx}{}^*(j\omega)}{P_F(j\omega)} - v_x(j\omega)\right\|_2.\end{aligned} \quad (4.34)$$

where the H_2 norm at the right-hand side of (4.34) can be expressed as:

$$\begin{aligned}\left\|\frac{i_{cx}{}^*(j\omega)}{P_F(j\omega)} - v_x(j\omega)\right\|_2 &= \sqrt{\left|v_x(j\omega_1) - \frac{i_{cx}{}^*(j\omega_1)}{P_F(j\omega_1)}\right|^2 + \sum_{n=2}^{h}\left|\frac{i_{cx}{}^*(j\omega_n)}{P_F(j\omega_n)}\right|^2} \\ &= \sqrt{\left|V_{dcf}\right|^2 + \sum_{n=2}^{h}\left|V_{dcn}\right|^2} \\ &= V_{dcm}.\end{aligned} \quad (4.35)$$

Substituting (4.35) into (4.34) and combing (4.30), we obtain:

$$\|T_x(j\omega)\|_\infty \le \frac{V_{dc}}{V_{dcm}}. \quad (4.36)$$

Under minimum DC-link voltage design, we should satisfy:

$$\|T_x(j\omega)\|_\infty = \max|T_x(j\omega)| \le 1. \quad (4.37)$$

From (4.37), to ensure the *LC*-HAPF compensating performance under minimum DC-link voltage condition, it requires the maximum magnitude of $|T_x(j\omega)| \le 1$, which can be easily observed and analyzed from the bode plot of $T_x(j\omega)$ as shown in Fig. 4.8.

Intuitively speaking, $\|T_x(j\omega)\|_\infty > 1$ yields a large overshoot under a step change of the system reference signal [16]. In this case, the controller requires a larger inverter voltage (or V_{dc}) to stabilize the *LC*-HAPF system, which would lead to overmodulation and degrades the compensation performance under minimum DC-link voltage operation.

4.5.2 Closed-Loop Response Analysis of Different Current Controllers

To verify the performance of the proposed VPIC controller for the *LC*-HAPF, a comparative study of the closed-loop response of the *LC*-HAPF with different controllers will be conducted out in this chapter. In the following, the control block diagrams and the parameter design for the QPRC [18, 19], PMRC [20–22], and SOSMC [23] are briefly introduced and discussed.

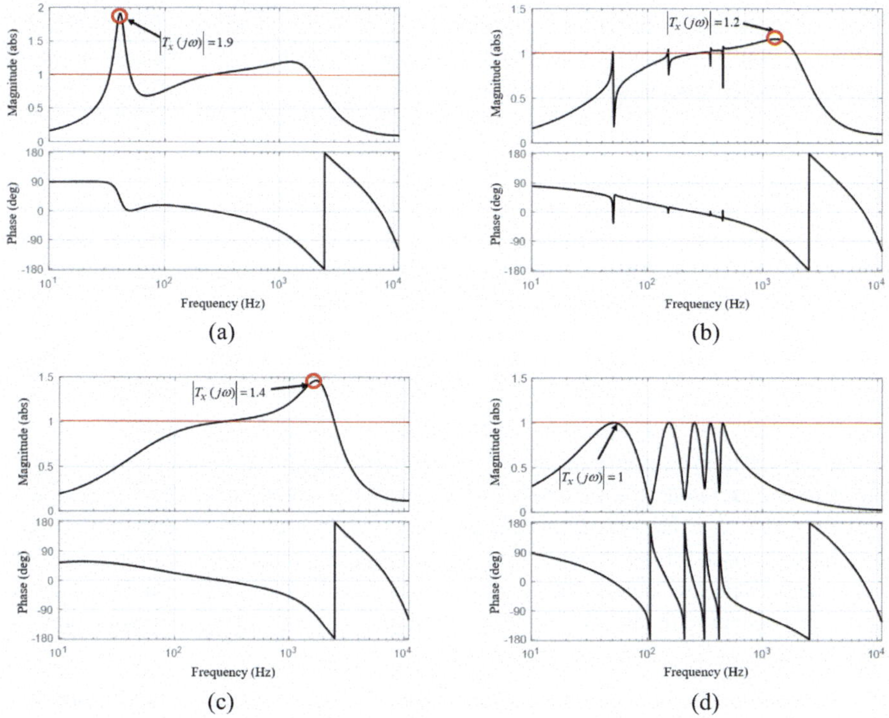

Fig. 4.8 Closed-loop responses of the *LC*-HAPF with different current controllers: **a** QPRC; **b** PMRC; **c** SOSMC; **d** VPIC

The *s*-domain control block diagram of different current controllers can also be represented by Fig. 3.1, where $C_{yx}(s)$ represents the transfer function of those controllers. It is worth noting that, although the SOSMC is a nonlinear controller, it can also be written as a PIC with variable control gains [23]. The system and control parameters used for the comparative study are given in Tables 4.1 and 4.2. The closed-loop responses of different current controllers are presented in Fig. 4.8.

(A) ***Quasi-Proportional-Resonant Controller***: The transfer function of the QPRC is presented as [18]:

$$C_{QPR}(s) = K_p + \frac{2K_r\omega_c s}{s^2 + 2\omega_c s + \omega_1^2}. \tag{4.38}$$

Table 4.2 Control parameters of different controllers

Controller	QPRC	PMRC	SOSMC	VPIC
Parameter	$K_p = 75$	$K_p = 50$	$\lambda_x = 150$	$K_x = 360$
	$K_r = 4500$	$K_m = 628$	$\kappa_x = 11000$	

where K_p is the proportional gain, K_r is the resonant gain, and ω_c is the bandwidth of the resonant frequency. According to [18], the proportional gain can be designed by the following inequality:

$$K_p \leq \frac{8L_c}{3T_s}. \tag{4.39}$$

where T_s is sampling time. The resonant gain K_r is suggested to make the resonant peak above 40 dB [18]. The ω_c is designed based on the allowable frequency variation standard, which is derived by:

$$\omega_c = 2\pi f_\Delta. \tag{4.40}$$

where f_Δ is the standard limitation of the allowable frequency variation.

The closed-loop response of the *LC*-HAPF with the QPRC is shown in Fig. 4.8a. The zero steady-state error condition Eq. (4.13) is satisfied at the fundamental frequency (50 Hz). However, the closed-loop response at other frequency ranges has a large deviation from 1, which means significant steady-state error at harmonic frequencies. Besides, a closed-loop peak as large as 2.1 ($\|T_x(j\omega)\|_\infty > 1$) will yield oscillatory output with slow settling time, and poor compensating performance with minimum DC-link voltage operation.

(B) ***Proportional Multi-resonant Controller***: The transfer function of the PMRC is given as [20–22]:

$$C_{PMR}(s) = K_p + \sum_{n=1}^{h} K_{rn} \frac{\cos(\phi_n)s - \omega_n sin(\phi_n)}{s^2 + \omega_n^2}. \tag{4.41}$$

where K_p is the proportional gain, K_{rn} and ϕ_n are the resonant gain and time-delay compensation angle for n-th harmonic order, respectively. The proportional gain of PMR decides the total bandwidth, which is suggested as [20–22]:

$$\left|K_p \cdot P_F(j\pi f_s/5)\right| \leq 1. \tag{4.42}$$

where f_s is the sampling frequency. The K_{rn} should be tuned to provide an adequate compromise between selective filtering and robustness under frequency variation, which is designed by:

$$K_{rn} = \beta_n K_p. \tag{4.43}$$

where β_n is the desired resonant bandwidth, typically hundreds of radians per second [20–22].

Figure 4.8b shows the closed-loop response of the *LC*-HAPF with the PMRC. The zero steady-state condition (4.13) is satisfied due to the infinite open-loop gain provided by the multiple resonant parts of (4.41). The multiple second-order integrators in the resonant part have a narrow passband around the harmonic frequencies. Thus, the harmonic transient response is slow. Besides, the closed-loop peak is up to 1.2 ($\|T_x(j\omega)\|_\infty > 1$), this will degrade the *LC*-HAPF compensating performance with minimum DC-link voltage operation.

(c) ***Second-Order Sliding Mode Controller***: The formula of the super-twisting based SOSMC [23] is given as:

$$v_{invx}{}^{*} = \lambda_x\sqrt{|\sigma_x|}\mathrm{sign}(\sigma_x) + \int \kappa_x \mathrm{sign}(\sigma_x). \quad (4.44)$$

where λ_x and κ_x are the control parameters of the SOSMC, and the compensating current error Δi_{cx} is considered as the sliding variable σ_x, i.e. $\sigma_x = \Delta i_{cx}$.

According to the small-signal analysis in [23], the SOSMC can be considered as a nonlinear PI controller, which has the following transfer function:

$$C_{SOSM}(s) = K_{hp}(\sigma_x) + \frac{K_{hi}(\sigma_x)}{s}. \quad (4.45)$$

where the proportional gain K_{hp} and integral gain K_{hi} are functions of the error signal:

$$K_{hp}(\Delta i_{cx}) = \lambda_x \frac{\sqrt{|\sigma_x|}\mathrm{sign}(\sigma_x)}{\sigma_x}. \quad (4.46)$$

$$K_{hi}(\Delta i_{cx}) = \kappa_x \frac{\mathrm{sign}(\sigma_x)}{\sigma_x}. \quad (4.47)$$

To guarantee the *LC*-HAPF system stability, robustness under parameter variation and disturbance rejection ability under distorted grid voltage of the SOSMC, the control parameters λ_x and κ_x should satisfy (4.48) and (4.49) [24]:

$$\kappa_x \frac{V_{dc}}{2L_c} > M_x. \quad (4.48)$$

$$\lambda_x \frac{V_{dc}}{2L_c} > \sqrt{\kappa_x \frac{V_{dc}}{2L_c} + M_x}. \quad (4.49)$$

where the auxiliary parameter M_x should be designed by [23]:

$$M_x = \max\left\{\left|I_{x1}\omega_1^2 - \frac{1}{L_c}V_{x1}\omega_1\right|\right\} + \max\left\{\left|\sum_{n=2}^{h}\frac{1}{L_c}V_{xn}(n\omega_1)^2 + \sum_{n=2}^{h}I_{xn}(n\omega_1)^2\right|\right\}. \quad (4.50)$$

where I_{x1} and I_{xn} are the magnitudes of the fundamental and nth order harmonic compensating currents; V_{x1} is the magnitude of the fundamental load voltage; V_{xn} is the magnitude of the n-th order harmonic load voltage, ω_1 is the fundamental angler frequency, n represents the harmonic order, h is the maximum harmonic order under consideration.

Figure 4.8c presents the closed-loop response of the *LC*-HAPF with the SOSMC. As the controllability analysis is based on the maximum magnitude $i_{cx}{}^*$ in (4.34) and (4.35), the worst-case of Δi_{cx} (6A for the nonlinear load in Table 4.1) is considered here. Due to the large control gains as shown in Table 4.2, the *LC*-HAPF with SOSMC also has a large closed-loop peak as 1.4 ($\|T_x(j\omega)\|_\infty > 1$), this will also degrade the *LC*-HAPF compensating performance with minimum DC-link voltage operation.

(D) ***Vector PI Controller***: Based on the transfer function and the optimal parameter design in 4.3, the closed-loop response of the *LC*-HAPF with VPIC is demonstrated in Fig. 4.8d. From Fig. 4.8d, the magnitude and phase at the considered harmonic order is $1 \angle 0°$, thus the zero steady-state error condition is verified. In addition, the bandwidth around each harmonic order is much larger than the PMRC case, which can obtain a faster and more accurate response for the harmonic signals. The closed-loop response is <1 at all frequency ranges. Thus, overshoot can be avoided, which optimizes the settling time [8]. Moreover, the controllability condition (4.37) is satisfied, which means *LC*-HAPF can obtain good compensating performance with minimum DC-link voltage.

4.6 Simulation Results

To verify the analytical results presented in the previous sections and the performance of the proposed VPIC, a MATLAB/Simulink based simulation are carried out. The system and load parameters are shown in Table 4.1. The testing loads are nonlinear rectifier loads, which are composed of three single-phase full-bridge rectifiers. Moreover, the control parameters of the different controllers are given in Table 4.2.

For verifying the validity and effectiveness of the proposed GPVI controller and its optimal parameter design method in 4.3, the simulated compensating performances of the VPIC controlled *LC*-HAPF system with the control gain K_x changes from 250 to 450 under 50 and 30 V DC-link voltages are plotted in Fig. 4.9, where the data is collected from the simulation results with K_x changes with a step of 10, the THDi_{sa} is the total harmonic distortion of phase a system current, T_{st} is the settling time.

From Fig. 4.9a, the compensated THDi_{sa} can be kept approximately the same with different K_x. This is because the variation of K_x only leads to a change of system bandwidth, which does not affect the *LC*-HAPF steady-state compensation performance as the zero steady-state error condition (4.14) is satisfied. As higher DC-link voltage leads to larger switching ripple, the THDi_{sa} with $V_{dc} = 50$ V is higher than that with $V_{dc} = 30$ V.

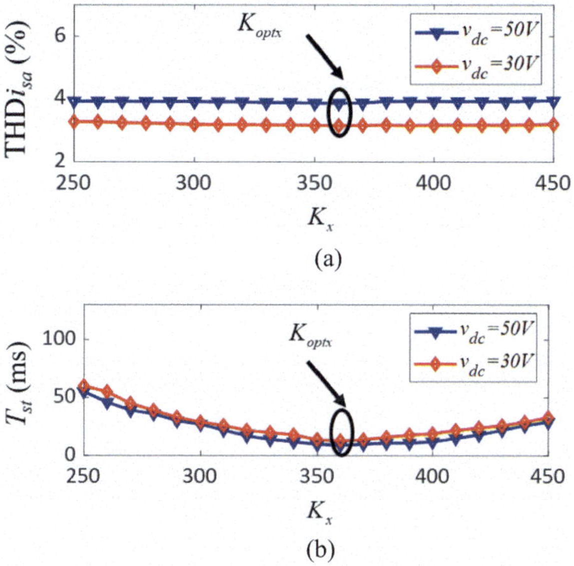

Fig. 4.9 Simulation results of the *LC*-HAPF system with the proposed VPIC with K_x changes from 250 → 450 under 50 V and minimum (30 V) DC-link voltages: **a** system current THD; **b** settling time

From Fig. 4.9b, the *LC*-HAPF system obtains the shortest settling time T_{st} with the optimal control gain K_{optx} for both $V_{dc} = 50$ V and $V_{dc} = 30$ V cases, which verifies the optimal parameter design in 4.3. In conclusion, the proposed VPIC and its parameter design method provide an optimal transient response speed and lowest THDi_{sa} simultaneously under 50 V and minimum (30 V) DC-link voltage cases.

To verify the robustness study of the *LC*-HAPF with VPIC in comparison with the other controllers, the simulation results of the *LC*-HAPF under the filtering parameter L_c and $C_c \pm 10\%$ variation are provided in Fig. 4.10. From Fig. 4.10a and c, the QPR controller has a weak robustness under system parameter variation. This is because there is only a resonant peak at the fundamental frequency, the harmonic compensation capability is mainly provided by the LC resonant peak. Once the resonant peak shift from the tuned frequency, the harmonic compensation performance will be degraded. For the other controllers, all of them provides ideally infinite control gain at all of the selected harmonic orders, thus shows a good robustness on harmonic compensation.

From Fig. 4.10b, the QPRC and PMRC also shows a poor system robustness on transient response speed when L_c varying. This is due to the transient response is mainly decided by the proportional gains; thus, the system bandwidth changes proportionally with the L_c. As is discussed in [23] and 4.3, the SOSMC and VPIC shows a good robustness on transient response speed. From Fig. 4.10d, all the controllers demonstrate a good robustness under C_c change, as C_c change only effect the system response in low frequency range.

In conclusion, when the L_c or C_c varies from −10 to +10% of the nominal value with the step of 1%, the *LC*-HAPF compensation performances are approximately the

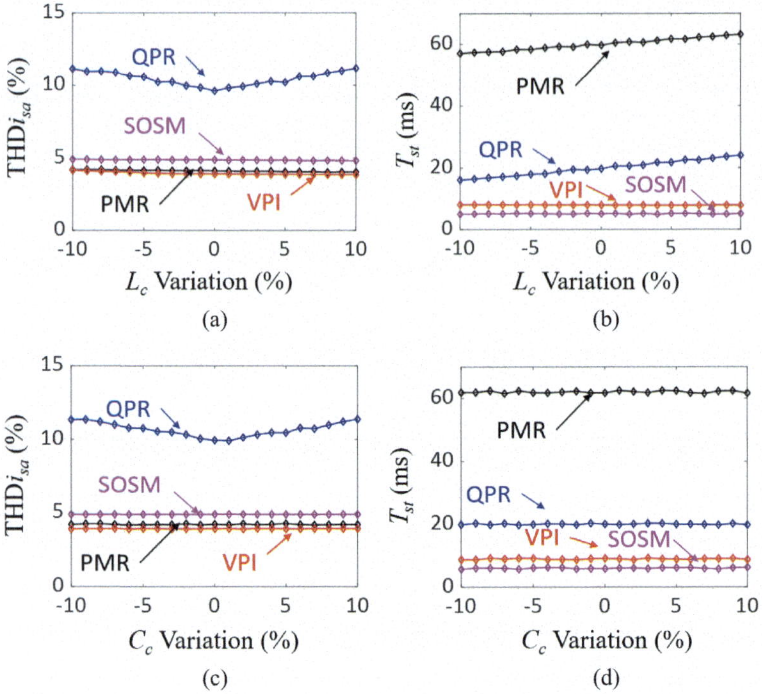

Fig. 4.10 Simulation results of the *LC*-HAPF with the different controllers under L_C and C_C variation: **a** system current THD with L_C variation ($-10 \rightarrow 10\%$); **b** settling time with L_C variation ($-10 \rightarrow 10\%$); **c** system current THD with C_C variation ($-10 \rightarrow 10\%$); **d** settling time with C_C variation ($-10 \rightarrow 10\%$)

same, which verifies the good system robustness of the proposed VPIC in comparison with the QPRC, PMRC, and SOSMC.

4.7 Experimental Results

To verify the analytical results presented in the previous sections and the performance of the proposed VPIC, a 110V-5kVA three-phase four-wire center-split *LC*-HAPF based experiment are carried out. The experimental prototype is built in the laboratory as discussed in Chap. 2. The system and load parameters are shown in Table 4.1 and the control parameters of the different controllers are given in Table 4.2.

The experimental results of the system current and load voltage waveforms before and after the *LC*-HAPF compensation under different DC-link voltages ($V_{dc} = 50$ and 30 V) and different current controllers are presented in Figs. 4.11 and 4.13, respectively. Their corresponding harmonic spectrums of system current after compensation are given in Figs. 4.12 and 4.14. The experimental transient and steady-state compensation performances of the *LC*-HAPF with different DC-link voltages and different current controllers are summarized in Table 4.3.

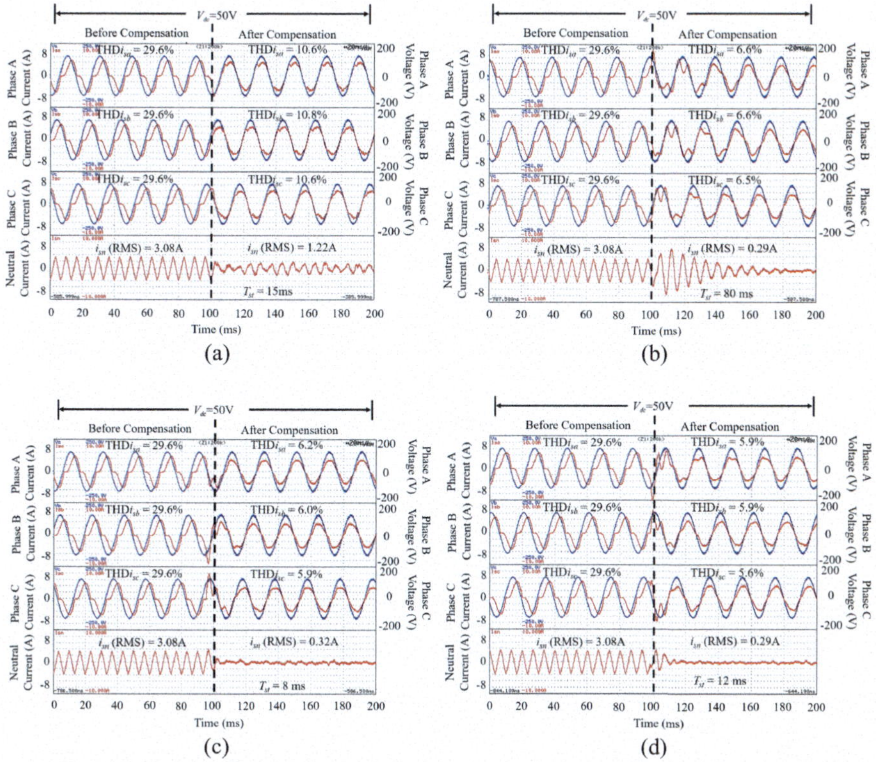

Fig. 4.11 Experimental results of system current and load voltage before and after *LC*-HAPF compensation under DC-link voltage V_{dc} = 50 V with different current controllers: **a** QPRC; **b** PMRC; **c** SOSMC; **d** VPIC

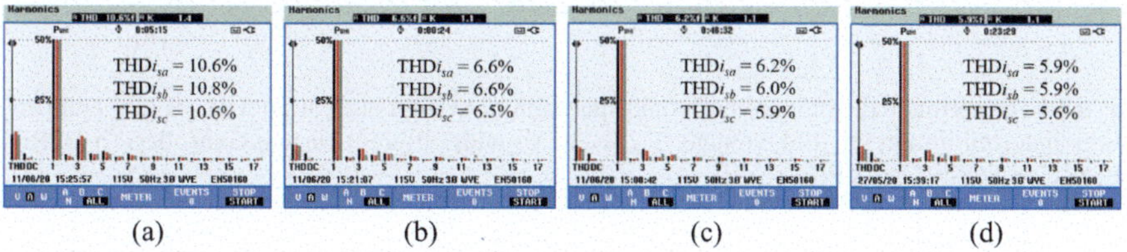

Fig. 4.12 Experimental results of harmonic spectrum of system current after *LC*-HAPF compensation under DC-link voltage V_{dc} = 50 V with different current controllers: **a** QPRC; **b** PMRC; **c** SOSMC; **d** VPIC

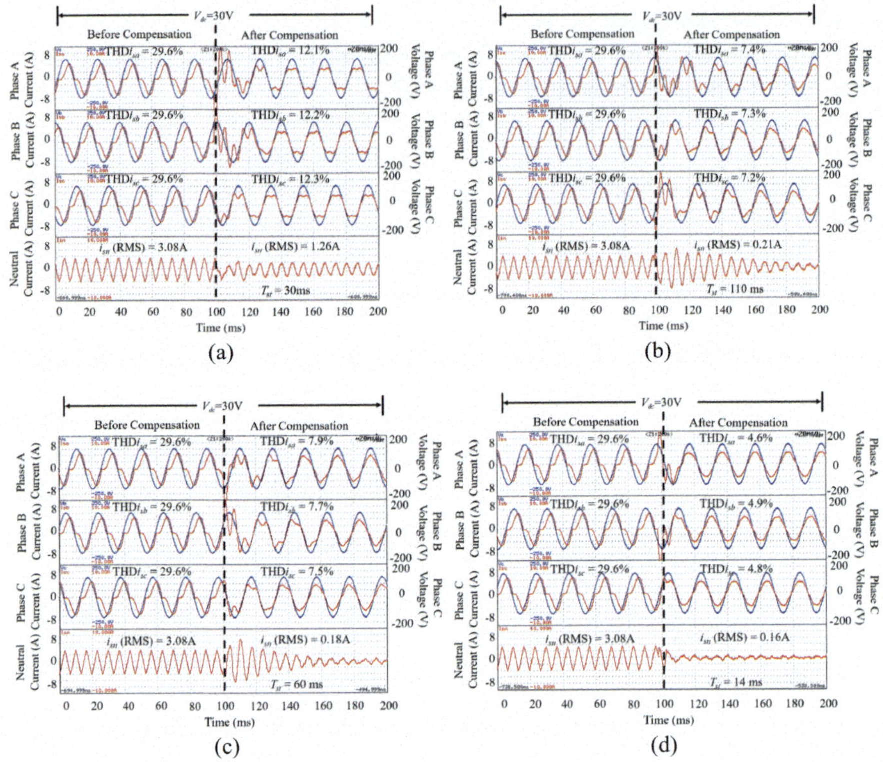

Fig. 4.13 Experimental results of system current and load voltage before and after *LC*-HAPF compensation under minimum DC-link voltage (V_{dc} = 30 V) with different current controllers: **a** QPRC; **b** PMRC; **c** SOSMC; **d** VPIC

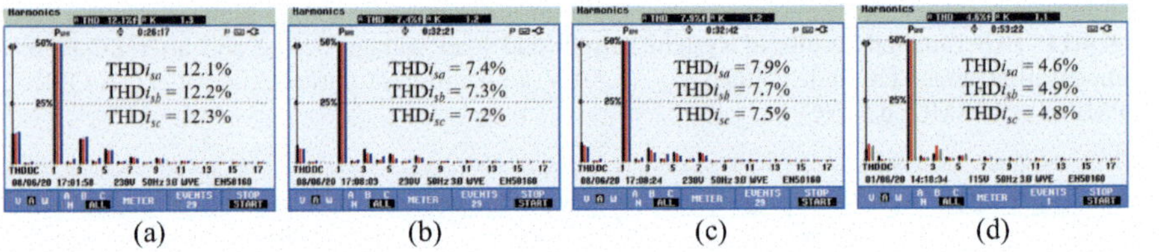

Fig. 4.14 Experimental results of harmonic spectrum of system current after *LC*-HAPF compensation under minimum DC-link voltage (V_{dc} = 30 V) with different current controllers: **a** QPRC; **b** PMRC; **c** SOSMC; **d** VPIC

(A) ***Quasi-Proportional-Resonant Controller***: From Figs. 4.11a and 4.13a, the *LC*-HAPF with the QPRC is stabilized with fast transient response. This is due to the relatively high proportional and resonant gains of the QPRC. However, as the resonant peak of the QPR controller is only tuned at the fundamental frequency, significant steady-state error occurs at other harmonic orders, which

Table 4.3 Experimental compensation results by *LC*-HAPF with different controllers and DC-link voltage

DC-link Voltage	Performance Index (phase *a*)	Before Compensation	After Compensation			
			QPRC	PMRC	SOSMC	VPIC
50 V	T_{st} (ms)	–	15	80	8	12
	PF	0.75	0.99	0.99	0.99	0.99
	THDi_{sx} (%)	29.6	10.6	6.6	6.2	5.9
	i_{sn} (Arms)	2.74	1.22	0.29	0.32	0.29
30 V	T_{st} (ms)	–	30	110	60	14
	PF	0.75	0.99	0.99	0.99	0.99
	THDi_{sx} (%)	29.6	12.1	7.4	7.9	4.6
	i_{sn} (Arms)	2.74	1.26	0.21	0.18	0.16

makes the compensated THDi_{sx} > 10%, as shown in Figs. 4.12 and 4.14. As shown in Fig. 4.8a, the closed-loop peak is up to 2.1 (requires V_{dc} > 63 V), thus condition (4.37) cannot be satisfied with both V_{dc} = 50 V and V_{dc} = 30 V. For the smaller V_{dc} = 30 V, the QPR controller provides a larger current distortion and longer settling time.

(B) ***Proportional Multi-Resonant Controller***: Figs. 4.11b and 4.13b present the dynamic and steady-state compensation performance of the *LC*-HAPF with the PMRC under V_{dc} = 50 V. The steady-state compensation performance is much better than that of the QPR controller, which is due to the multiple resonant parts that provide infinite control gain at the selected current harmonic orders. However, refer to Fig. 4.11b, the settling time T_{st} is up to 80 ms. This is because the integrator characteristic of its resonant parts leads to a narrow bandwidth for harmonics. From Fig. 4.8b, the closed-loop peak is 1.2 (requires v_{dc} > 36 V), which means the PMR controller works well with V_{dc} = 50 V, as shown in Fig. 4.11b. But the compensation performance will be degraded under the minimum DC-link voltage V_{dc} = 30 V. As shown in Fig. 4.13b, the transient response speed becomes slower (T_{st} = 110 ms).

(C) ***Second-Order Sliding Mode Controller***: The SOSMC provides a fast transient response (T_{st} = 8 ms) and a low THDi_{sa} under DC-link voltage V_{dc} = 50 V, as shown in Figs. 4.11c and 4.13c. However, as shown in Fig. 4.8c, the closed-loop peak of 1.4 (requires V_{dc} > 42 V) makes the SOSMC degrade the compensating performance with the minimum DC-link voltage V_{dc} = 30 V. The control performance is significantly degraded (7.5 times slower) with V_{dc} = 30 V, as shown in Figs. 4.11c and 4.13c.

(D) ***Vector PI controller***: The experimental results of the proposed VPIC under v_{dc} = 50 and 30 V cases are demonstrated in Figs. 4.11d, 4.12d, 4.13d and 4.14d. The transient response speed by using the VPIC is fast with T_{st} = 12 and 14 ms, respectively, which is comparable with the SOSMC under V_{dc} = 50 V. In the meanwhile, the compensated THDi_{sx} can be lower than 6%, which is the best among the controllers. Under V_{dc} = 30 V, the proposed VPIC provides the

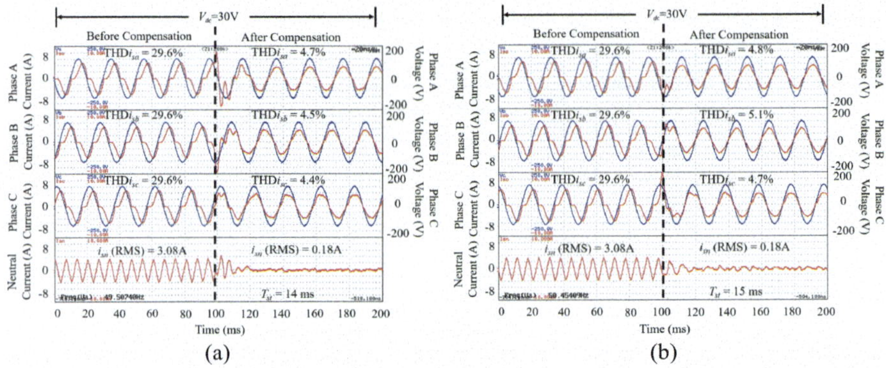

Fig. 4.15 Experimental results of system current and load voltage before and after the VPIC controlled *LC*-HAPF compensation under different grid frequency variation: **a** 49.5 Hz; **b** 50.5 Hz

fastest transient response speed and lowest THDi_{sa} (THDi_{sx} < 5%) among the controllers. Overall, the proposed VPIC obtains the lowest THDi_{sa} performance and fast transient response under 50 V and minimum (30 V) DC-link voltage conditions.

To verify the robustness of *LC*-HAPF with VPIC under grid frequency variation, the experimental results of the VPIC controlled *LC*-HAPF under 49.5, 50 and 50.5 Hz are provided in Fig. 4.15. From Fig. 4.15, the proposed VPIC shows a good robustness under the worst-case grid frequency variation defined by the IEEE Std. 446-1995 [9].

To verify the robustness of *LC*-HAPF with VPIC under grid voltage distortion, the experimental results of the *LC*-HAPF with VPIC under grid harmonic voltage with 8.1% THD (contains 3% of 5th harmonic order component), which is the worst-case defined by IEEE STD 519-2014 [25] of voltage THD <8%, are demonstrated in Fig. 4.16, in which the system current THDi_{sx} can be compensated to 5.2% from 29.6% with a settling time of 8 ms, which clearly illustrates that the *LC*-HAPF with the proposed VPIC obtains good system robustness under gird voltage distortion situation.

For the unbalanced nonlinear load situation within the *LC*-HAPF compensation range shown in Table 4.1, the experimental result of the *LC*-HAPF with the proposed VPIC is shown in Fig. 4.17. Figure 4.17 verifies that it can compensate the unbalanced nonlinear loading situation provided that the unbalanced loading is within the *LC*-HAPF compensation range. The unbalance factor (UF) of the three-phase system currents is reduced from 13.7 to 1.8%.

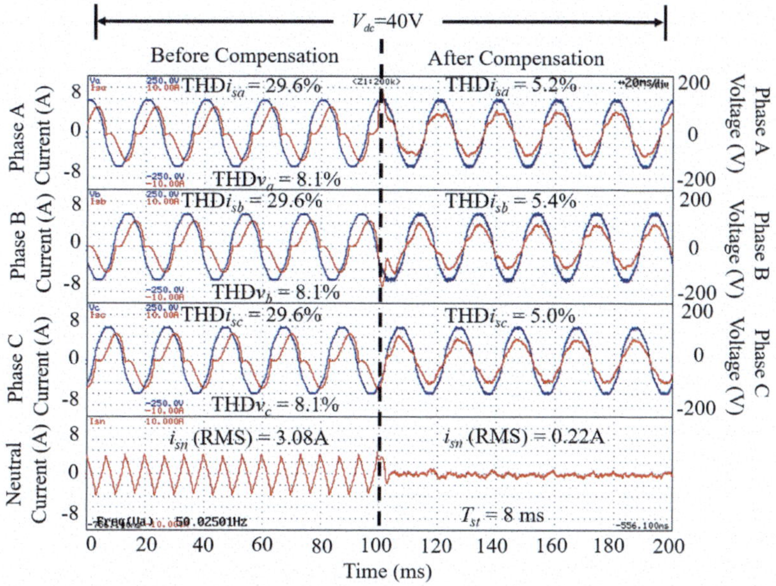

Fig. 4.16 Experimental results of load voltage and system current compensated by the *LC*-HAPF with VPIC under distorted grid voltage

4.8 Summary

A VPIC for the *LC*-HAPF with ideally zero steady-state error, fast transient response, and good robustness under its minimum DC-link voltage condition is developed and presented in this chapter. The transfer function of the VPIC is proposed for the *LC*-HAPF that ensures ideally zero steady-state error, and the parameter design of the VPIC for optimal transient response speed is also presented. Moreover, a detailed robustness analysis of the VPIC including the system parameter variation, distorted grid voltage and grid frequency variation conditions is proposed. Then, the digital implementation method of the VPIC and an overall control block diagram for the *LC*-HAPF with the proposed VPIC are also proposed. To highlight the advantages of the proposed VPIC, a comparative study of the VPIC with other representative current controllers under minimum DC-link voltage condition is analyzed and discussed, in which the VPIC obtains the best performances. To verify the comparative study, the performances of the proposed VPIC for the *LC*-HAPF is verified by representative simulation and experimental results in comparison with the QPRC, PMRC and SOSMC controllers, achieving the best steady-state compensation performance, and fast transient response under both sufficient and minimum DC-link voltage conditions. Finally, the robustness under different grid conditions and the unbalance compensation capability of the proposed VPIC is also verified by experimental results.

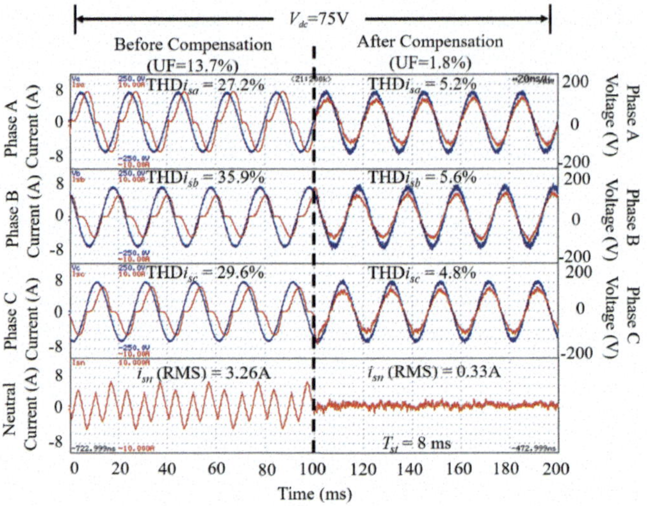

(a)

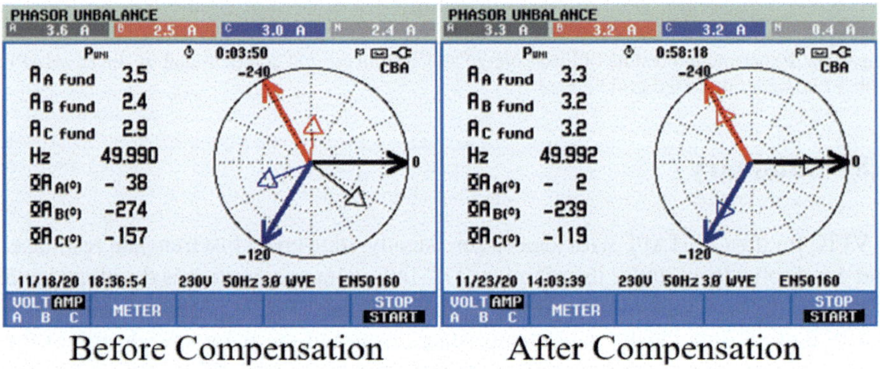

(b)

Fig. 4.17 Experimental results of load voltage and system current before and after compensation compensated by *LC*-HAPF with VPIC for unbalanced nonlinear load compensation: **a** load voltage and system current waveforms; **b** load voltage and system current phasors

References

1. S. Buso, P. Mattavelli, *Digital Control in Power Electronics* (Morgan & Claypool Publishers, 2015)
2. J. Sun, Small-signal methods for ac distributed power systems—a review. IEEE Trans. Power Electron. **24**(11), 2545–2554 (2009)
3. C.-S. Lam, X.-X. Cui, W.-H. Choi, M.-C. Wong, Y.-D. Han, Minimum inverter capacity design for LC-hybrid active power filters in three-phase four-wire distribution systems. IET Power Electron. **5**(7), 956–968 (2012)

4. C.-S. Lam, W.-H. Choi, M.-C. Wong, Y.-D. Han, Adaptive DC-link voltage-controlled hybrid active power filters for reactive power compensation. IEEE Trans. Power Electron. **27**(4), 1758–1772 (2012)
5. C.-S. Lam, M.-C. Wong, W.-H. Choi, X.-X. Cui, H.-M. Mei, J.-Z. Liu, Design and performance of an adaptive low-DC-voltage-controlled LC-hybrid active power filter with a neutral inductor in three-phase four-wire power systems. IEEE Trans. Ind. Electron. **61**(6), 2635–2647 (2014)
6. G.C. Goodwin, S.F. Graebe, M.E. Salgado, and others, *Control System Design*, vol. 240 (Prentice Hall New Jersey, 2001)
7. A.G. Yepes, F.D. Freijedo, J. Doval-Gandoy, Ó. López, J. Malvar, P. Fernandez-Comesaña, Effects of discretization methods on the performance of resonant controllers. IEEE Trans. Power Electron. **25**(7), 1692–1712 (2010)
8. A.G. Yepes, A. Vidal, J. Malvar, O. Lopez, J. Doval-Gandoy, Tuning method aimed at optimized settling time and overshoot for synchronous proportional-integral current control in electric machines. IEEE Trans. Power Electron. **29**(6), 3041–3054 (2014)
9. IEEE Recommended Practice for Emergency and Standby Power Systems for Industrial and Commercial Applications. *IEEE Std 446-1995 Orange Book* (1996), pp. 1–320
10. R. Bojoi, L.R. Limongi, F. Profumo, D. Roiu, A. Tenconi, Analysis of current controllers for active power filters using selective harmonic compensation schemes. IEEJ Trans. Electr. Electron. Eng. **4**(2), 139–157 (2009)
11. A.G. Yepes, F.D. Freijedo, Ó. Lopez, J. Doval-Gandoy, High-performance digital resonant controllers implemented with two integrators. IEEE Trans. Power Electron. **26**(2), 563–576 (2011)
12. L. Harnefors, Implementation of resonant controllers and filters in fixed-point arithmetic. IEEE Trans. Ind. Electron. **56**(4), 1273–1281 (2008)
13. F. Rodriguez, E. Bueno, M. Aredes, L. Rolim, F.A. Neves, M.C. Cavalcanti, Discrete-time implementation of second order generalized integrators for grid converters, in *2008 34th Annual Conference of IEEE Industrial Electronics* (2008), pp. 176–181
14. H. Akagi, E.H. Watanabe, M. Aredes, *Instantaneous Power Theory and Applications to Power Conditioning* (Wiley, 2017)
15. W.-H. Choi, C.-S. Lam, M.-C. Wong, Y.-D. Han, Analysis of DC-link voltage controls in three-phase four-wire hybrid active power filters. IEEE Trans. Power Electron. **28**(5), 2180–2191 (2013)
16. S. Skogestad, I. Postlethwaite, *Multivariable Feedback Control: Analysis and Design*, vol. 2 (Wiley, New York, 2007)
17. K. Deimling, *Nonlinear Functional Analysis* (Courier Corporation, 2010)
18. T. Ye, N. Dai, C.-S. Lam, M.-C. Wong, J.M. Guerrero, Analysis, design, and implementation of a quasi-proportional-resonant controller for a multifunctional capacitive-coupling grid-connected inverter. IEEE Trans. Ind. Appl. **52**(5), 4269–4280 (2016)
19. T. Ye, N. Dai, K.W. Lao, C.-S. Lam, M.-C. Wong, Low DC voltage PV generation system with power factor correction and harmonic suppression capability in a distribution network. IET Gener. Transm. Distrib. **13**(7), 1049–1056 (2019)
20. Z. He, F. Ma, M. Li, Q. Xu, F. Zhou, Y. Yue, Multi-functional hybrid active power converter and its industrial application for electrolytic copper-foil. IET Gener. Transm. Distrib. **10**(16), 3985–3994 (2016)
21. A. Tan, K.Ç. Bayındır, M.U. Cuma, M. Tümay, Multiple harmonic elimination-based feedback controller for shunt hybrid active power filter. IET Power Electron. **10**(8), 945–956 (2017)
22. C. Xie, X. Zhao, K. Li, J. Zou, J.M. Guerrero, A new tuning method of multiresonant current controllers for grid-connected voltage source converters. IEEE Trans. Emerg. Sel. Top. Power. Electron. **7**(1), 458–466 (2019)

23. C. Gong, W.-K. Sou, C.-S. Lam, Second-order sliding-mode current controller for LC-coupling hybrid active power filter. IEEE Trans. Ind. Electron. **68**(3), 1883–1894 (2021)
24. R. Seeber, M. Horn, Stability proof for a well-established super-twisting parameter setting. Automatica **84**, 241–243 (2017)
25. IEEE Standard 519-2014, *IEEE Recommended Practices and Requirements for Harmonic Control in Electrical Power Systems* (2014)

Chapter 5
Second Order Sliding Model Control for *LC*-HAPF

5.1 Introduction

In this chapter, a second-order sliding mode control (SOSMC) based on super-twisting controller for *LC*-HAPF is proposed. Among the literatures review, hysteresis based current controller, which is a special implementation of sliding mode control, is usually used due to its simplicity, robustness, and good dynamic performance for reference tracking. However, it generates a non-fixed switching frequency for the power filter which leads to undesirable harmonics problem. The proposed SOSMC design ensures the good robustness and dynamic performance of using the conventional SMC with an improved steady-state performance and disturbance rejection capability. In addition, with the help of super-twisting controller, the proposed SOSMC design can achieve fixed switching frequency characteristic. First, the control objective and the design methodology of sliding surface are discussed and given in this chapter. Second, a detailed parameter design procedure under stability, system parameter variation, distorted grid voltage and grid frequency variation consideration is provided. Then, experimental results on a three-phase four-wire *LC*-HAPF are carried out to verify the steady-state and dynamic performance of the proposed SOSMC design compared with other representative current controllers. Finally, the disturbance rejection capability under different grid conditions of the proposed SOSMC design is also verified by experimental results.

The contribution of this chapter is summarized as follows:

(1) Propose a SOSMC current control strategy for *LC*-HAPF with low steady-state error, fast transient response, fixed switching frequency, good robustness, and grid disturbance rejection capability.
(2) Study the stability, robustness, disturbance rejection, and the parameter design of the SOSMC for the *LC*-HAPF to ensure its stability and performance under parameter variation, distorted grid voltage, and grid frequency variation.
(3) Propose an overall control block diagram for the *LC*-HAPF with SOSMC.

C.-S. Lam et al., *Advanced Controllers Design for Hybrid Active Power Filters*,
https://doi.org/10.1007/978-981-97-6847-9_5

(4) Verify the proposed SOSMC design for the *LC*-HAPF by simulation and experimental results in comparison with the conventional HCC, the QPRC, and the PMRC.

5.2 Control Objective and Super-Twisting Controller

5.2.1 Control Objective

From the circuit configuration of the three-phase four-wire *LC*-HAPF as shown in Fig. 1.4, a single-phase equivalent diagram of the *LC*-HAPF with a reactive and harmonic current detection block, a current controller and a PWM block are shown in Fig. 5.1, where $i_{cx}{}^{*}$ represents reference current, $v_{invx}{}^{*}$ is the reference inverter voltage, u_x is the output of current controller. For simplicity, the grid voltage is assumed to balanced. The dynamic model of the single-phase equivalent system can be derived as follows:

$$L_c \frac{di_{cx}}{dt} = v_x - v_{cx} - \frac{V_{dc}}{2} u_x. \tag{5.1}$$

$$C_c \frac{dv_{cx}}{dt} = i_{cx} \tag{5.2}$$

The main idea of the SOSMC is to guarantee the system trajectory converges to the second-order sliding surface and keeps sliding on it. The motion on the second-order sliding surface is called the second-order sliding mode control [1]. Hence, the first step of the SOSMC design is to find out the sliding surface, which is considered as the control objective.

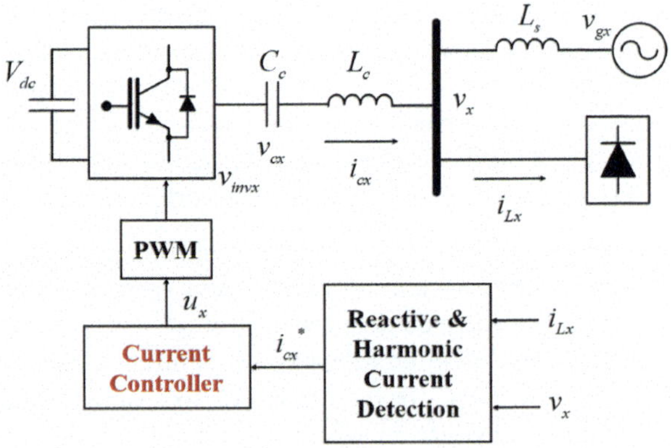

Fig. 5.1 Single-phase equivalent diagram of the *LC*-HAPF

For the *LC*-HAPF, the control objective of the proposed SOSMC design is to generate an actual compensating current accurately equals to the reference compensating current calculated by the reactive and harmonic current detection block (compensating current error equals to 0), which can be expressed as:

$$i_{cx}{}^{*} - i_{cx}=0 \tag{5.3}$$

where $i_{cx}{}^{*}$ and i_{cx} are the reference and actual compensating currents. Follow this control objective for the proposed SOSMC design, the sliding variable σ_x [1] can be defined as (5.4), note that $\sigma_x = \Delta i_{cx}$.

$$\sigma_x = i_{cx}{}^{*} - i_{cx} \tag{5.4}$$

The proposed SOSMC design needs to drive the sliding variable σ_x to 0 to minimize the current error. At the same time, the time derivative of the sliding variable σ_x should be equaled to 0 as well, which prevents the compensating current oscillation at steady-state. Thus, the sidling surface of the proposed SOSMC design is provided as [1]:

$$\sigma_x=\frac{d\sigma_x}{dt} = 0 \tag{5.5}$$

Under (5.5) condition, the current error and its rate of change are eliminated, which refers to a good steady-state performance without high frequency oscillation.

5.2.2 *Super-Twisting Controller*

After the choice of sliding surface, the second step is to design the control law that makes the sliding variable σ and its time derivative converge to the designated sliding surface and keep sliding on it. There are many control laws for SOSMC [1–3], but most of them provide a non-continuous output *u*, and a real-time measurement of $d\sigma_x/dt$ is needed. To avoid these drawbacks, the super-twisting controller [4] that provides a continuous output u_x without $d\sigma_x/dt$ detection is used.

The formula of the super-twisting controller [4] is given as:

$$u_x = u_{x1} + u_{x2} \tag{5.6}$$

$$u_{x1} = \lambda_x\sqrt{|\sigma_x|}\text{sign}(\sigma_x) \tag{5.7}$$

$$\frac{du_{2x}}{dt} = \begin{cases} \kappa_x\text{sign}(\sigma_x) & |u_x| \le U_{Mx} \\ -u_x & |u_x| > U_{Mx} \end{cases} \tag{5.8}$$

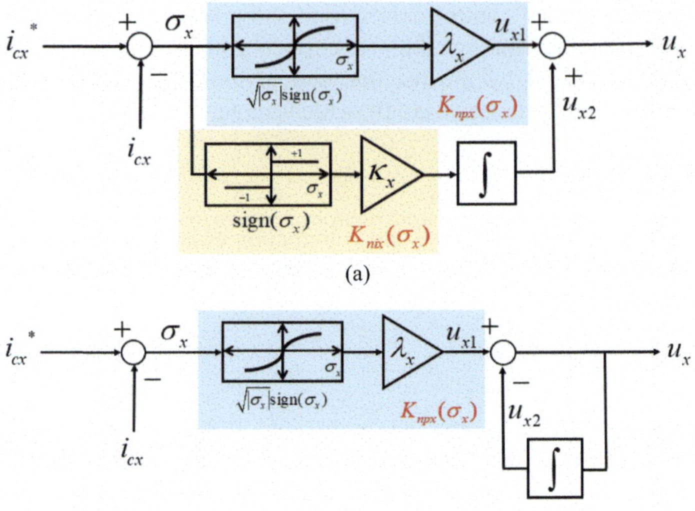

Fig. 5.2 Block diagram of the proposed SOSMC design with super-twisting controller: **a** normal condition; **b** saturated condition

where λ_x and κ_x are the control gains of the SOSMC, U_{Mx} is the peak value of triangle carrier wave for PWM generation.

The block diagram of the SOSMC with super-twisting controller under normal and saturated condition is given in Fig. 5.2. Note that the high frequency switching term sign(σ_x) is smoothed by the integrator. Therefore, a continuous control output u_x can be obtained by the SOSMC under super-twisting controller. After that, u_x is used to compare with a fixed frequency carrier wave to generate the trigger signals for the *LC*-HAPF, thus achieving fixed switching frequency even under SOSMC.

In the case $|u_x| \leq U_{Mx}$, the controller is operating at normal state as shown in Fig. 5.2a. The block diagram can be considered as a nonlinear version of the conventional PI controller. The feedforward path from σ_x to u_{x1} is a nonlinear proportional gain $K_{npx}(\sigma_x)$ that is changing adaptively for different σ_x, where $K_{npx}(\sigma_x)$ is expressed as:

$$K_{npx}(\sigma_x) = \frac{u_{x1}}{\sigma_x} = \lambda_x \frac{\sqrt{|\sigma_x|}\text{sign}(\sigma_x)}{\sigma_x} \tag{5.9}$$

From (5.9), $K_{npx}(\sigma_x)$ provides a proportional gain smaller than λ_x for $|\sigma_x|>1$ while a proportional gain larger than λ_x for $|\sigma_x|<1$. For example, when $\sigma_x = 100$ or -100 ($|\sigma_x|>1$), $u_{x1} = 10\lambda_x$ via (5.7), thus the proportional gain $K_{npx}(\sigma_x) = 0.1\lambda_x$ via (5.9). Usually, σ_x is large during heavy load changes, the controller will adaptively reduce the proportional gain $K_{npx}(\sigma_x)$ value in order to improve the system stability range for large transient disturbance. On the contrary, when $\sigma_x = +0.01$ or $\sigma_x = -0.01$ ($|\sigma_x|<1$), $K_{npx}(\sigma_x) = 10\lambda_x$ is obtained via (5.9). When the sliding variable σ

(compensating current error) becomes smaller, the proportional gain $K_{npx}(\sigma_x)$ will become larger to further reduce the system steady-state error. In steady-state, σ_x is very small ($\sigma_x \approx 0$), the proportional gain $K_{npx}(\sigma_x)$ is sufficiently large, which implies better steady-state performance than any fixed linear proportional gain.

The other feedforward path from σ_x to u_{x2} can be considered as a nonlinear integral gain $K_{nix}(\sigma_x)$, which is expressed as:

$$K_{nix}(\sigma_x) = \frac{du_{x2}}{dt}\frac{1}{\sigma_x} = \kappa_x \frac{\text{sign}(\sigma_x)}{\sigma_x} \tag{5.10}$$

Similar to $K_{npx}(\sigma_x)$, $K_{nix}(\sigma_x)$ also provides an integral gain smaller than α for $|\sigma_x|>1$ while an integral gain larger than α for $|\sigma_x| < 1$. As an example, when $\sigma_x =$ 100 or -100 ($|\sigma_x|>1$), $du_{x2}/dt = +\kappa_x$ or $-\kappa_x$ via (5.8), but they yield the same $K_{nix}(\sigma_x)$ $= 0.01\kappa_x$ via (5.10). During heavy load changes (σ_x is large), a small integral gain represents a narrow system bandwidth, which reduces the system sensitivity to large transient disturbance. On the contrary, when $\sigma_x = +0.01$ or $\sigma_x = -0.01$ ($|\sigma_x|<1$), $du_{x2}/dt = +\kappa_x$ or $-\kappa_x$, and $K_{nix}(\sigma_x)$ becomes 100κ. The nonlinear integral gain $K_{nix}(\sigma_x)$ also tends to infinitely large when σ_x is very small ($\sigma_x \approx 0$) at steady-state, thus the steady-state error is further reduced.

For $|u_x|>U_{Mx}$, the controller is operating at saturated state as shown in Fig. 5.2b. In this case, u_{x2} provides negative integral feedback of u_x to reduce $|u_x|$ back to the normal state, hence over-modulation can be effectively prevented.

5.3 Design of Second Order Sliding Model Control for *LC*-HAPF

5.3.1 Under System Stability Study

The final step is to design the control gains, λ_x and κ_x of the SOSMC. From the system model and the super-twisting controller (5.1), (5.2), (5.4), (5.6)–(5.8), with the help of [5], a canonical form of the *LC*-HAPF system of SOSMC with the super-twisting controller is deduced and given as:

$$\frac{d\sigma_x}{dt} = \lambda_x \frac{v_{dc}}{2L_c}\sqrt{|\sigma_x|}\text{sign}(\sigma_x) + \gamma_x \tag{5.11}$$

$$\frac{d\gamma_x}{dt} = \kappa_x \frac{v_{dc}}{2L_c}\text{sign}(\sigma_x) + \delta_x \tag{5.12}$$

where γ_x is an auxiliary variable, δ_x stands for the total 'perturbation' of the system (5.11), (5.12) which can be written as:

$$\delta_x = \frac{d^2 i_{cx}^*}{dt^2} - \frac{1}{L_c}\left(\frac{dv_x}{dt} - \frac{i_{cx}}{C_c}\right) \tag{5.13}$$

Assuming δ_x is bounded by:

$$|\delta_x| \le M_x \tag{5.14}$$

where M_x is a constant and the upper boundary of δ_x. The system (5.11)–(5.13) is stable against the perturbation (5.14) if its designed parameters satisfy (5.15) and (5.16) [5]:

$$\kappa_x \frac{v_{dc}}{2L_c} > M_x \tag{5.15}$$

$$\lambda_x \frac{v_{dc}}{2L_c} > \sqrt{\kappa_x \frac{v_{dc}}{2L_c} + M_x} \tag{5.16}$$

The detailed stability proof of (5.15) and (5.16) can be found in [5]. However, how to obtain the value of M_x from the system parameters and external conditions was missing in [5], which makes the stability design in [5] cannot be directly applied to stability study of the *LC*-HAPF.

In practical case, the *LC*-HAPF system performances may also suffer from system parameter variation, distorted grid voltage, and grid frequency variation. However, the robustness and disturbance rejection capability of the SOSMC controlled system are still lack in study among the existing literatures [1–5]. To address these concerns, this section proposes an analytical form of M_x describing its relationship with system parameters and grid conditions. Moreover, a parameter design guideline for the SOSMC controlled *LC*-HAPF is proposed in this chapter for achieving good stability, robustness, and disturbance rejection capability.

From (5.14), Mx can be obtained based on the worst-case consideration of δ_x. Under the worst-case consideration, when there is no compensating current in (5.13), it becomes:

$$|\delta_x| = \left| \frac{d^2 i_{cx}^*}{dt^2} - \frac{1}{L_c}\left(\frac{dv_x}{dt} - \frac{i_{cx}}{C_c}\right) \right| \le \left| \frac{d^2 i_{cx}^*}{dt^2} - \frac{1}{L_c}\frac{dv_x}{dt} \right| \tag{5.17}$$

However, i_{cx}^* and v_x in (5.17) are time-varying. To find the constant M_x, the waveforms of i_{cx}^* and v_x should be known. Under inductive loading compensation and distorted grid voltage consideration, i_{cx}^* contains inductive reactive and harmonic components of the load current. Then the load voltage v_x and the reference compensating current i_{cx}^* can be expressed as:

$$v_x = V_{x1}\cos(\omega_1 t + \varphi_{vx1}) + \sum_{n=2}^{h} V_{xn}\sin(h\omega_1 t + \varphi_{vxn}) \tag{5.18}$$

$$i_{cx}{}^{*} = -I_{x1}\sin(\omega_1 t + \varphi_{ix1} - \frac{\pi}{2}) - \sum_{n=2}^{h} I_{xn}\sin(h\omega_1 t + \varphi_{ixn}) \tag{5.19}$$

where I_{x1} and I_{xn} are the magnitude of the fundamental reactive and nth harmonic compensating currents; V_{x1} is the magnitude of the fundamental load voltages; V_{xn} is the magnitude of the harmonic load voltages. ω_1 is the fundamental angular frequency, n represents the harmonic order, h is the maximum harmonic order under consideration, φ_{vx1}, φ_{vxn}, and φ_{ix1}, φ_{ixn} are the phase of v_x at fundamental and harmonic order, and phase of $i_{cx}{}^{*}$ at fundamental and harmonic order, respectively.

Substituting (5.18) and (5.19) into (5.17), the following inequality can be obtained:

$$\begin{aligned}\left|\frac{d^2 i_{cx}{}^{*}}{dt^2} - \frac{1}{L_c}\frac{dv_x}{dt}\right| &= \left|\begin{aligned}&(I_{x1}\omega_1^2 - \frac{1}{L_c}V_{x1}\omega_1)\cos(\omega_1 t + \varphi_{vx1})\\ &+\sum_{n=2}^{h}\frac{1}{L_c}V_{xn}(h\omega_1)\sin(h\omega_1 t + \varphi_{vxn})\\ &+\sum_{n=2}^{h} I_{xn}(h\omega_1)^2\sin(h\omega_1 t + \varphi_{ixn})\end{aligned}\right|\\ &\le \left|I_{x1}\omega_1^2 - \frac{1}{L_c}V_{x1}\omega_1\right| + \left|\sum_{n=2}^{h}\frac{1}{L_c}V_{xn}(h\omega_1) + \sum_{n=2}^{h} I_{xn}(h\omega_1)^2\right|\end{aligned} \tag{5.20}$$

Consequently, the analytical form of M_x can be derived as:

$$M_x = \max\left\{\left|I_{x1}\omega_1^2 - \frac{1}{L_c}V_{x1}\omega_1\right|\right\} + \max\left\{\left|\sum_{n=2}^{h}\frac{1}{L_c}V_{xn}(h\omega_1)^2 + \sum_{n=2}^{h} I_{xn}(h\omega_1)^2\right|\right\} \tag{5.21}$$

In order to achieve good stability, robustness and disturbance rejection capability of the *LC*-HAPF with SOSMC, the M_x should be designed based on (5.21), in which (5.21) can be separated into two parts: the maximum value of subtraction part and the maximum value of summation part.

In the following, the parameter design guideline for the stable SOSMC controlled *LC*-HAPF achieving good robustness and disturbance rejection capability will be proposed based on three main considerations: parameter variation, grid voltage distortion and grid frequency distortion.

5.3.2 Under System Parameter Variation Consideration

Under parameter variation of the *LC*-HAPF, from (5.21), the M_x has a relationship with the L_c value, in which the variation range of L_c can be written as:

$$L_c \in \Delta_{L_c} = [(1 - z_{L_c})L_c, (1 + z_{L_c})L_c] \tag{5.22}$$

where z_{Lc} is the maximum variation percentage, and ΔL_c is a set containing all possible values of the L_c. From (5.21), the M_x value under parameter variation defined by (5.22) can be derived as:

$$M_x|_{L_c} = \max\{M_x[(1 - z_{L_c})L_c], M_x[(1 + z_{L_c})L_c]\} \tag{5.23}$$

In order to satisfy the stability and parameter variation consideration of the *LC*-HAPF with SOSMC, the M_x should be designed based on (5.23).

5.3.3 Under Distorted Grid Voltage Consideration

The range of load harmonic voltage magnitude at each harmonic order is limited by IEEE or other standards, which can be written as:

$$V_{xn} \in \Delta_{V_{xn}} = [0, V_{xn}{}^M], \, for\, n = 2, 3, \ldots, h \tag{5.24}$$

where $V_{xn}{}^M$ is the maximum magnitude of the grid voltage harmonics at each harmonic order *n*, and ΔV_{xn} is a set containing all possible values of the V_{xn} at each harmonic order. From (5.21), the M_x value under distorted grid voltage defined by (5.24) can be derived as:

$$M_x|_{V_{xn}} = M_x(V_{x2}{}^M, V_{x2}{}^M, \ldots, V_{xn}{}^M) \tag{5.25}$$

In order to satisfy the stability and distorted grid voltage consideration of the *LC*-HAPF with SOSMC, the M_x should be designed based on (5.25).

5.3.4 Under Grid Frequency Variation Consideration

The range of the grid frequency variation is limited by IEEE or other standards, which can be written as:

$$\omega_1 \in \Delta_{\omega_1} = [(1 - z_{\omega_1})\omega_1, (1 + z_{\omega_1})\omega_1] \tag{5.26}$$

where $z_{\omega 1}$ is the variation percentage of the grid frequency, $\Delta\omega_1$ is a set containing all possible values of the ω_1. From (5.21), the M_x value under grid frequency variation defined by (5.26) can be derived as:

$$M_x|_{\omega_1} = \max\{M_x[(1 - z_{\omega_1})\omega_1], M_x[(1 + z_{\omega_1})\omega_1]\} \tag{5.27}$$

In order to satisfy the stability and grid frequency variation consideration of the *LC*-HAPF with SOSMC, the M_x should be designed based on (5.27).

5.3.5 *Final M_x Design Under Parameter Variation, Distorted Grid Voltage and Grid Frequency Variation Consideration*

For the parameter variation and grid frequency variation bounded by (5.22) and (5.26), the maximum value of subtraction part in (5.21) can be derived as:

$$\max\left\{\left|I_{x1}\omega_1^2 - \frac{1}{L_c}V_{x1}\omega_1\right|\right\} = \max\left\{\begin{array}{l} M_x[(1 - z_{L_c})L_c, (1 - z_{\omega_1})\omega_1], \\ M_x[(1 + z_{L_c})L_c, (1 - z_{\omega_1})\omega_1], \\ M_x[(1 - z_{L_c})L_c, (1 + z_{\omega_1})\omega_1], \\ M_x[(1 + z_{L_c})L_c, (1 + z_{\omega_1})\omega_1] \end{array}\right\} \tag{5.28}$$

Based on (5.28), the maximum value of subtraction part in (5.21) can be obtained. The maximum value of summation part in (5.21) can be calculated by substituting all of the ${V_{xn}}^M$ in (5.24) into (5.21). Adding these two parts together, the final M_x in (5.21) can be obtained and finally the control gains λ_x and κ_x can be determined by (5.15) and (5.16).

In conclusion, the design guidelines of the proposed *LC*-HAPF with SOSMC are given as follows:

(1) Set the system parameters variation, grid voltage harmonics, and grid frequency variation as in (5.22), (5.24) and (5.26);
(2) Determine the constant M_x by (5.21) with the help of (5.28);
(3) Calculate the control gains λ_x and κ_x by (5.15) and (5.16);
(4) Build the super-twisting controller based SOSMC by (5.6)–(5.8).

5.4 Overall Control Block Diagram

The overall control block diagram of the *LC*-HAPF with SOSMC is provided in Fig. 5.3, which includes an instantaneous power compensation control block and a final reference compensating current SOSMC block. The instantaneous power

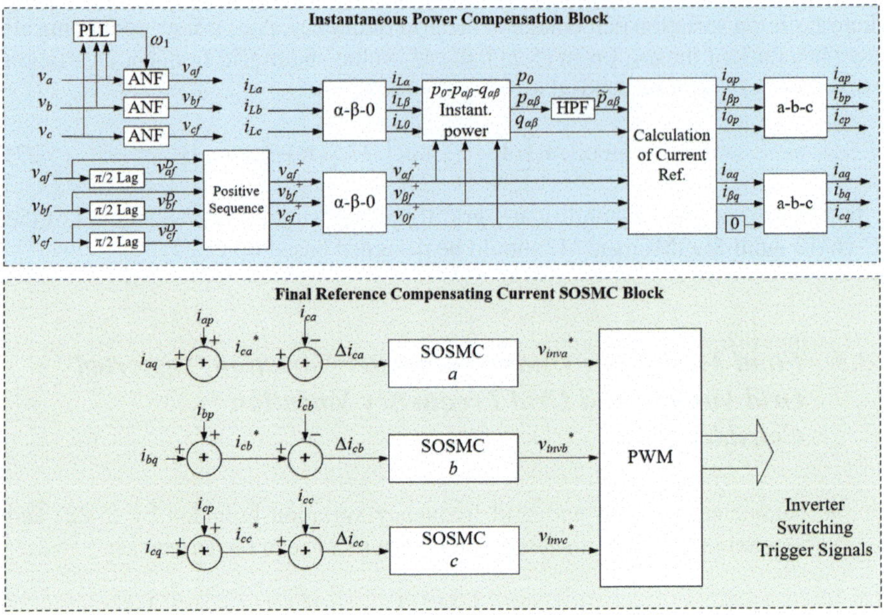

Fig. 5.3 Overall control block diagram of *LC*-HAPF with SOSMC

compensation control block is used for harmonic and reactive power detection based on the instantaneous *p*-*q* theory [6]. To relax the effect of the voltage harmonic to the reference compensating current calculation, a band-pass filter (BPF) based symmetrical component extraction algorithm [7] is used to extract fundamental positive-sequence components of the load voltage as the input voltage for the instantaneous *p*-*q* reference current calculation. For the final reference compensating current SOSMC block, the reference compensating current in each phase compares with the measured compensating current and provides the current error signal σ_x. Then, the current error signal of each phase feeds into the super-twisting controller and generates the control signals u_x, which is the instantaneous duty ratio of the PWM. Finally, the inverter switching trigger signals of the *LC*-HAPF are obtained from the PWM block. In this chapter, a fixed frequency carrier wave is used to compare with u_x in order to generate the inverter trigger signals.

5.5 Simulation Results

For verifying the performance of the proposed SOSMC design for the *LC*-HAPF, a MATLAB/Simulink based simulation is firstly carried out. The system and load parameters are presented in Table 5.1. The testing Load Set 1 are nonlinear loads, which are composed of three single-phase full-bridge rectifiers, while the Load Set

2 are composed of Load Set 1 and linear loads. Moreover, the control parameters of the different controllers are given in Table 5.2.

The SOSMC design in (5.6)–(5.8) already provides the lower bound design constraint of the control parameters, which guarantees the system stability, robustness, and disturbance rejection ability. However, the control gains in practical engineering cannot be designed to be extremely large, which may lead to saturation effect and control performance degradation. To address this concern, a sweeping-based design is used to find a locally optimal solution at the neighborhood of the lower bounds, and thus avoiding unnecessarily large control parameters.

The simulated compensating performances of the *LC*-HAPF system with SOSMC under various λ_x and κ_x parameters are plotted in Fig. 5.4, where the data is collected from the 100 times simulation results of the *LC*-HAPF with different λ_x and κ_x parameters. The λ_x varies from 100 to 200 with a step-size of 2 while the κ_x varies from 8000 to 12000 with a step-size 80.

From Fig. 5.5, the variation of control parameters of the SOSMC satisfies the parameter range provided by (5.15), (5.16) in (5.3). The variation of λ_x has a less significant effect on system performance comparing to κ_x, as shown in Fig. 5.5. The increasing of λ_x improves the compensating performances, but the improvement is not significant as $\lambda_x \geq 150$. Moreover, the settling time can keep <4.5 ms as $\lambda_x \leq 150$. For $\lambda_x > 150$, the settling time becomes longer. This is because too large λ_x gain leads to saturation and disables the integrator.

From Fig. 5.5, the increasing of κ_x improves the compensating performances, but the improvement is not significant as $\kappa_x \geq 11000$. Moreover, the optimum settling time is close to $\kappa_x = 11000$. For $\kappa_x > 11000$, the settling time becomes longer. This is because too large κ_x gain also leads to saturation and disables the integrator.

Table 5.1 System parameters for simulation and experiment

System Parameters	Physical Values
v_{gx}, f_g	110 V_{rms}, 50 Hz
L_S	0.5 mH
R_{NL}, L_{NL}, C_{NL} (Nonlinear rectifier load)	43 Ω, 35 mH, 400 μF
R_{LL}, L_{LL} (Linear load)	60 Ω, 70 mH
L_c, C_c	8 mH, 50 μF
V_{dc}	90 V
f_s	10 kHz
f_{sw}	10 kHz

Table 5.2 Control parameters of different controllers

Controller	QPRC	PMRC	HCC	SOSMC
Parameter	$K_p = 50$	$K_p = 20$	$H = 0.0781$	$\lambda_x = 150$
	$K_r = 6250$	K_{rn}=600		κ_x=11000

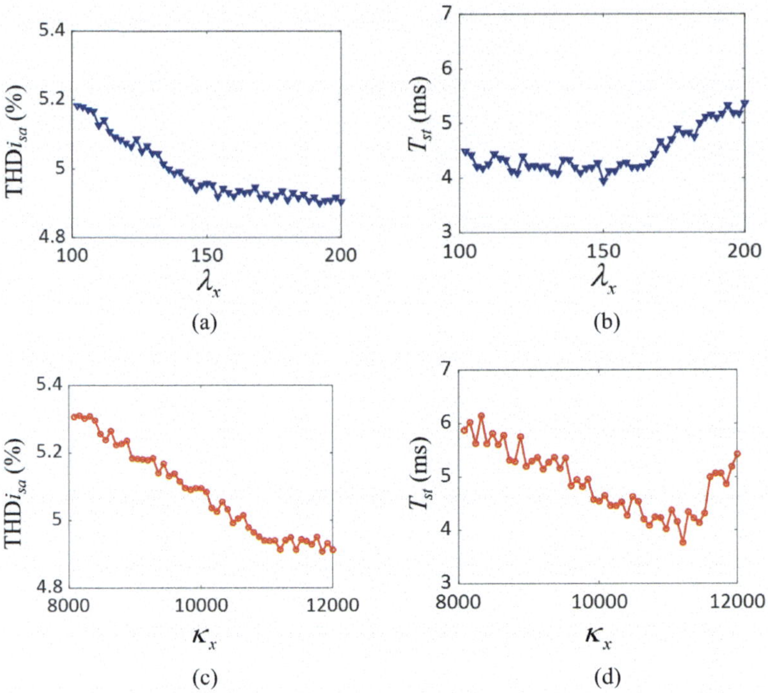

Fig. 5.4 Simulation results of the *LC*-HAPF system with SOSMC under various λ_x and κ_x parameters: **a** system current THD with $\kappa_x = 11000$, $\lambda_x = 100 \to 200$; **b** settling time with $\kappa_x = 11000$, $\lambda_x = 100 \to 200$; **c** system current THD with $\kappa_x = 8000 \to 12000$, $\lambda_x = 150$; **d** settling time with $\kappa_x = 8000 \to 12000$, $\lambda_x = 150$

To verify the robustness of the *LC*-HAPF, simulation results under the system parameter L_c and C_c variation of the *LC*-HAPF are carried out. The L_c and C_c vary from -10 to $+10\%$ of the nominal value with the step 0.5%. The simulation results under the parameter variation study are provided in Fig. 5.5, the compensation performances are approximately the same, which verifies the robustness of the designed *LC*-HAPF system with SOSMC.

5.6 Experimental Results

To verify the analytical results presented in the previous sections and the performance of the proposed SOSMC, a 110V-5kVA three-phase four-wire center-split *LC*-HAPF based experiment are carried out. The experimental prototype is built in the laboratory as discussed in Chap. 2.

The experimental results of the system current and voltage waveforms by *LC*-HAPF under different load condition with the proposed SOSMC design and with

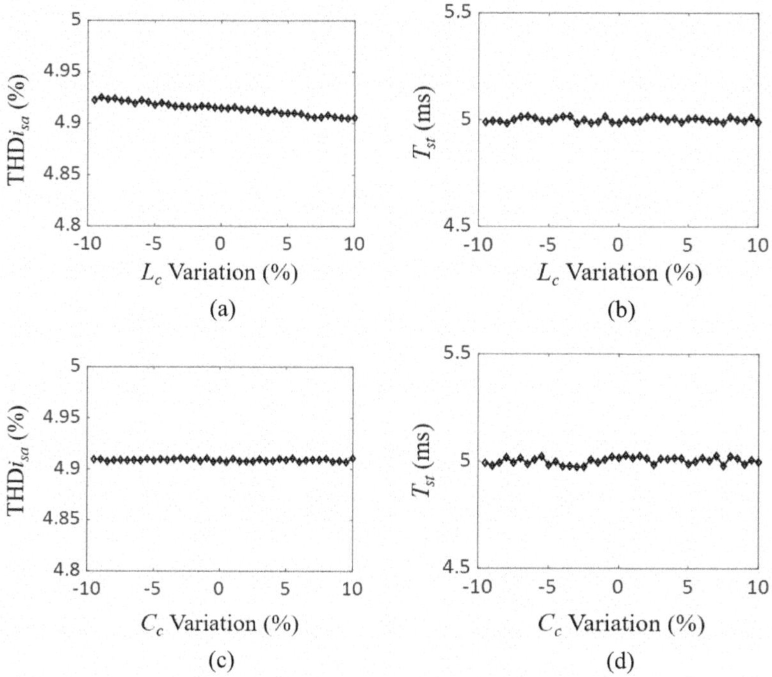

Fig. 5.5 Simulation results of the *LC*-HAPF system with SOSMC under L_C and C_C variation: **a** system current THD with L_C variation ($-10 \rightarrow 10\%$); **b** settling time with L_C variation ($-10 \rightarrow 10\%$); **c** system current THD with C_C variation ($-10 \rightarrow 10\%$); **d** settling time with C_C variation ($-10 \rightarrow 10\%$)

different controllers QPRC [8, 9], PMRC [10–13] and HCC [14, 15] are presented in Figs. 5.6 and 5.8. Their corresponding harmonic spectrums of compensated system current are given in Figs. 5.7 and 5.9. High order harmonic spectrum of the system current with Load Set 1 with the HCC and the proposed SOSMC design are demonstrated in Fig. 5.10. The transient and steady-state performance indexes of the experimental results with these controllers are summarized in Table 5.3.

From Figs. 5.6, 5.7, 5.8, 5.9 and 5.10 and Table 5.3, the experimental results of the QPRC, PMRC, HCC and the proposed SOSMC design can compensate the system power factor to 0.98 or above and meet the compensated source current THD $\leq 15\%$ (IEEE STD 519-2014 [17]) under both of the testing Load Set 1 and 2. Among them, only HCC and the proposed SOSMC design obtain the fast response speed of T_{st} <10 ms. Besides that, the PMRC and the proposed SOSMC design obtain the best performance on THD suppression.

From Fig. 5.10, the HCC has no high frequency ripple at 10 kHz (200th order) caused by fixed frequency PWM, but the harmonics at low-frequency range is serious. Besides, the experimental average switching frequency of HCC 3.1 kHz in for Load Set 1 and 4.7 kHz for Load Set 1. On the contrary, the switching frequency of SOSMC

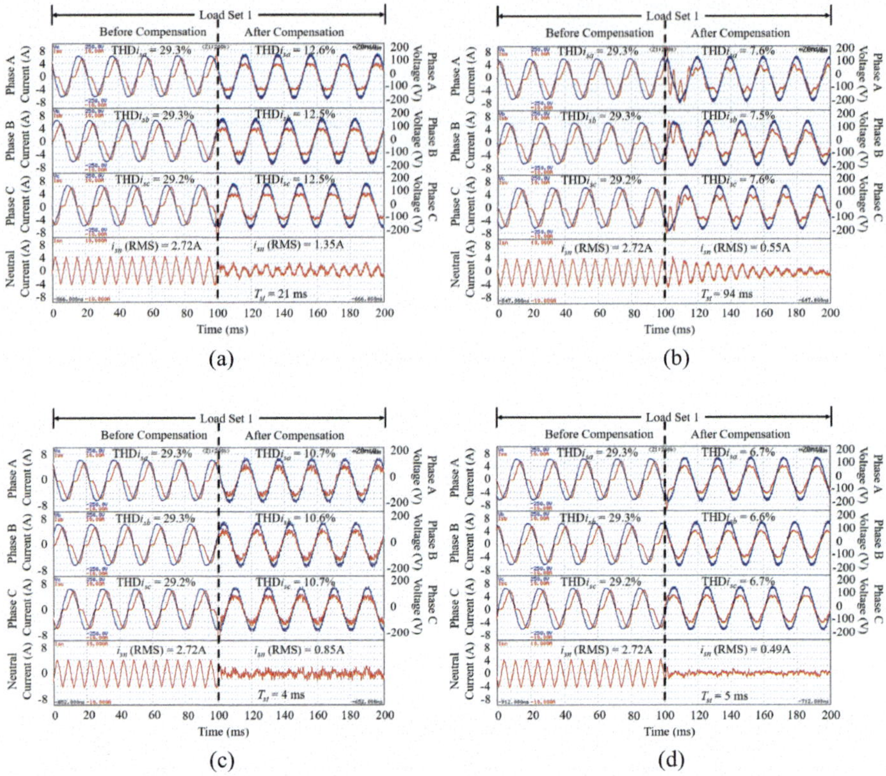

Fig. 5.6 Experimental results of load voltage and system current before and after compensation under Load Set 1 by *LC*-HAPF with different controllers: **a** with QPRC; **b** with PMRC; **c** with HCC; **d** with SOSMC

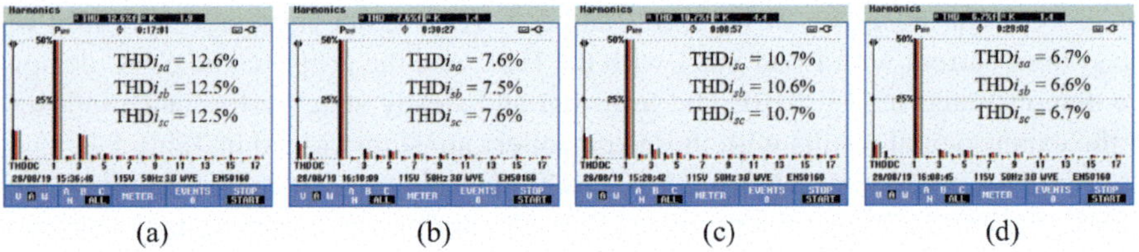

Fig. 5.7 Experimental results of harmonic spectrum of system current under Load Set 1 by *LC*-HAPF with different controllers: **a** with QPRC; **b** with PMRC; **c** with HCC; **d** with SOSMC

is fixed at 10 kHz and the low order harmonic decreases significantly compared with the HCC.

Due to the relatively high proportional, resonant gains and the simplicity of the controller, the QPRC obtains an acceptable transient performance as shown in Figs. 5.6a and 5.8a. However, the steady-state error is large, as shown in Figs. 5.7a

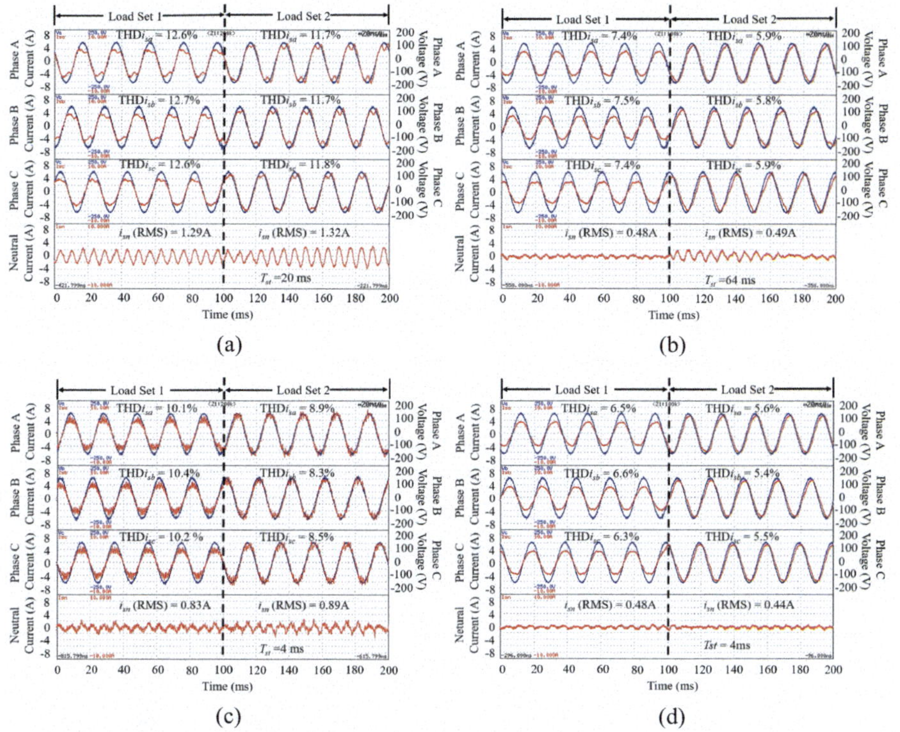

Fig. 5.8 Experimental results of load voltage and compensated system current under a step load change by *LC*-HAPF with different controllers: **a** with QPRC; **b** with PMRC; **c** with HCC; **d** with SOSMC

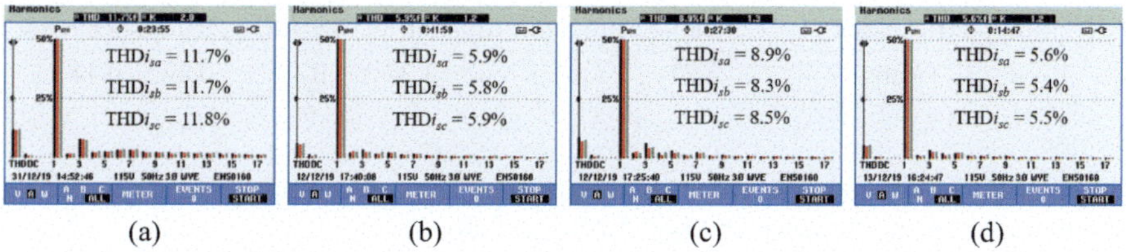

Fig. 5.9 Experimental results of harmonic spectrum of system current under Load Set 2 by *LC*-HAPF with different controllers: **a** with QPRC; **b** with PMRC; **c** with HCC; **d** with SOSMC

and 5.9a. This is because the QPRC is designed to have one open loop resonant peak at the fundamental frequency, while the LC filter provides another open loop resonant peak at the 5th harmonic only. Therefore, the QPRC cannot compensate well the other harmonic orders, especially for the 3rd harmonic order.

For the PMRC, as it provides multiple resonant peaks for the selected harmonic orders, it can achieve ideally zero steady-state error at those harmonic orders. Therefore, it can achieve very good steady-state performance, as shown in Figs. 5.7b

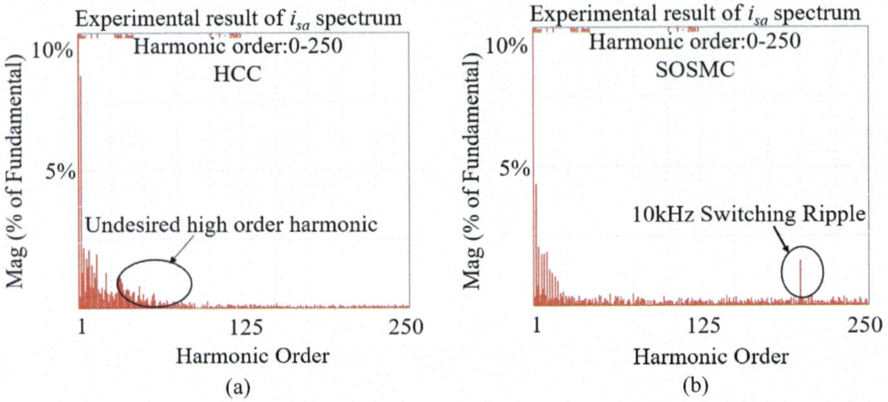

Fig. 5.10 Experimental results of harmonic spectrum of system phase a current compensated by *LC*-HAPF with varying and fixed frequency controllers: **a** with HCC; **b** with SOSMC

Table 5.3 Experimental compensation results for load set 1 and load set 2 by *LC*-HAPF with different controllers

Load	Performance Index (phase *a*)	Before Compensation	After Compensation			
			QPRC	PMRC	HCC	SOSMC
Load Set 1	T_{st} (ms)	–	21	94	4	5
	PF	0.75	0.99	0.99	0.98	0.99
	THDi_{sx} (%)	29.3	12.5	7.6	10.7	6.7
	i_{sn} (Arms)	2.72	1.28	0.55	0.85	0.49
Load Set 2	T_{st} (ms)	–	20	64	4	4
	PF	0.81	0.99	0.99	0.98	0.99
	THDi_{sx} (%)	21.3	11.7	5.9	8.9	5.6
	i_{sn} (Arms)	2.73	1.32	0.49	0.89	0.44

and 5.9b. However, the integrator characteristic of its resonant parts leads to a slow transient response, as shown in Figs. 5.6b and 5.8b.

According to the analysis results of the first-order sliding mode control [16], the HCC as a particular case of the conventional first-order sliding mode controller can obtain an ideally infinite proportional gain. Thus, it provides a very fast transient response, as shown in Figs. 5.6c and 5.8c. However, the hysteresis band modulation leads to a varying switching frequency, which generates undesired high-order harmonics that cannot be easily filtered out by a single-tuned output filter, as shown in Fig. 5.10a.

From the high-order sliding mode control theory [1], the super-twisting controller based SOSMC maintains the fast response speed property of the conventional SMC, as shown in Figs. 5.6d and 5.8d, with improved accuracy and fixed frequency characteristics, as shown in Figs. 5.7d, 5.8d and 5.10b. The reason of improving accuracy has been discussed by the small-signal method in Sect. 5.3, in which the SOSMC provides ideally infinite proportional and integral gains when the compensating error is small, thus leading a low steady-state error in a wide frequency range. Therefore, the transient response performance of the SOSMC is comparable to the HCC, and the steady-state performance is comparable to the PMRC, as shown in Table 5.3.

To verify the effectiveness of the proposed *LC*-HAPF with SOSMC under distorted grid voltage condition, Fig. 5.11 shows the experimental results of the *LC*-HAPF with SOSMC under distorted grid voltage condition (total harmonics distortion of the load voltage THDv_x = 11.5%, which does not satisfy the IEEE STD 519–2014 [17] of voltage THD < 8%), in which the system current THDi_{sx} can be compensated to 7.3% from 30.9% with a settling time of 7 ms, which satisfies the standard of current THD < 15% and verifies the *LC*-HAPF with SOSMC can perform satisfactory compensation even under distorted grid voltage condition.

To verify the effectiveness and robustness of the proposed controller under frequency variation, the experimental results of the *LC*-HAPF with SOSMC under frequency variation are provided. According to the IEEE Standard 446-1995 [18], the maximum frequency variation tolerance of the 50 Hz distribution power system is ±0.5 Hz. In this chapter, a more severe frequency variation of ±1 Hz is taken into consideration, and the corresponding experimental results are shown in Fig. 5.12. As the grid frequency variation has been taken consideration during the SOSMC design in Sect. 5.3, the experimental compensating performances of the *LC*-HAPF with

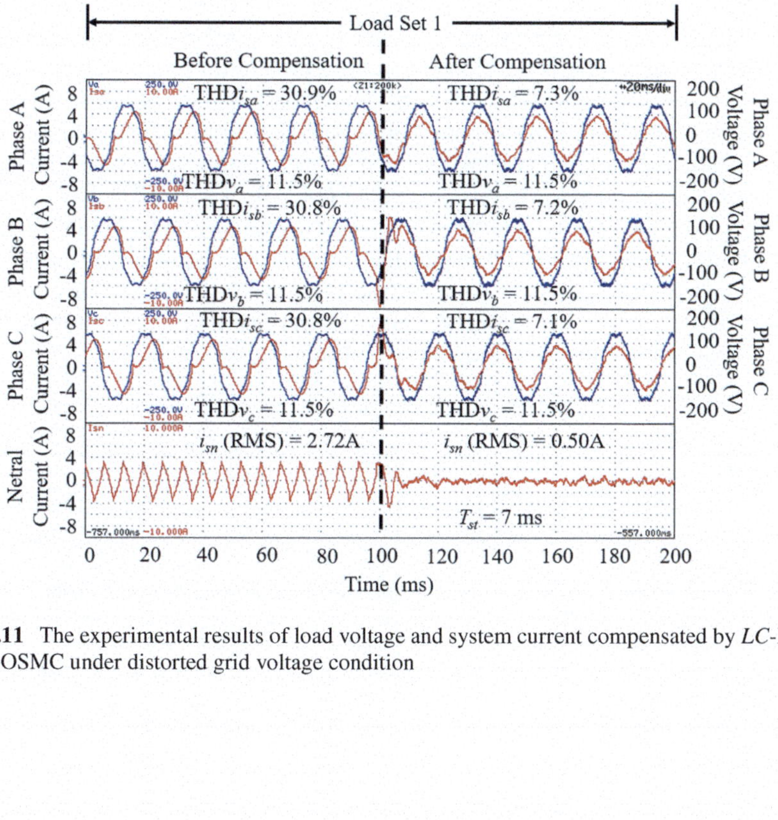

Fig. 5.11 The experimental results of load voltage and system current compensated by *LC*-HAPF with SOSMC under distorted grid voltage condition

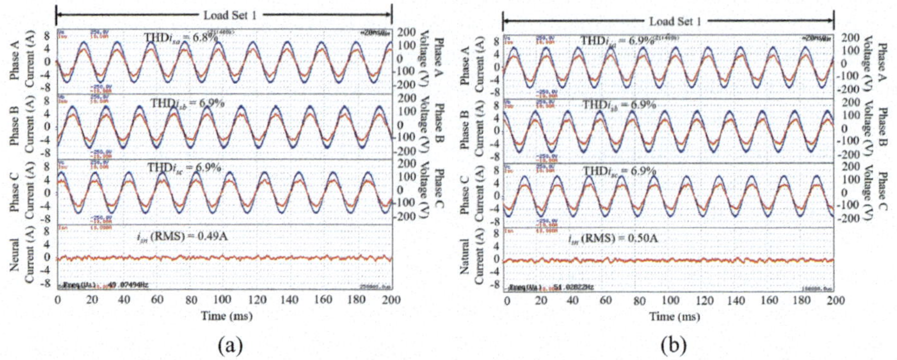

Fig. 5.12 The experimental results of load voltage and system current compensated by *LC*-HAPF with SOSMC under frequency variation: **a** under 49 Hz; **b** under 51 Hz

SOSMC controller are approximately the same even under ±1 Hz grid frequency variation.

5.7 Summary

A SOSMC current control strategy for the *LC*-HAPF with low steady-state error, fast transient response, fixed switching frequency, good robustness and grid disturbance rejection capability is developed and presented in this chapter. The stability, robustness, disturbance rejection and the parameter design of the SOSMC for the *LC*-HAPF under system parameter variation, grid voltage distortion, and grid frequency variation is studied and proposed with a step-by-step design guideline. The performances of the proposed SOSMC design for the *LC*-HAPF is verified by representative simulation and experimental results in comparison with the conventional HCC, the QPRC, and the PMRC, which shows a superior dynamic response and steady-state performance simultaneously. In addition, the robustness and disturbance rejection capability of the proposed SOSMC design is also verified under parameter variation and non-ideal grid voltage and frequency conditions.

References

1. Y. Shtessel, C. Edwards, L. Fridman, A. Levant, *Sliding Mode Control and Observation* (Springer, 2014)
2. A. Levant, Principles of 2-sliding mode design. Automatica. **43**(4), 576–586 (2007)
3. G. Bartolini, A. Pisano, E. Punta, E. Usai, A survey of applications of second-order sliding mode control to mechanical systems. Int. J. Control. **76**(9–10), 875–892 (2003)
4. A. Levant, Sliding order and sliding accuracy in sliding mode control. Int. J. Control. **58**(6), 1247–1263 (1993)

5. R. Seeber, M. Horn, Stability proof for a well-established super-twisting parameter setting. Automatica **84**, 241–243 (2017)
6. H. Akagi, E.H. Watanabe, M. Aredes, *Instantaneous Power Theory and Applications to Power Conditioning* (Wiley, 2017)
7. D. Yazdani, M. Mojiri, A. Bakhshai, G. Joós, A fast and accurate synchronization technique for extraction of symmetrical components. IEEE Trans. Power Electron. **24**(3), 674–684 (2009)
8. T. Ye, N. Dai, C.-S. Lam, M.-C. Wong, J.M. Guerrero, Analysis, design, and implementation of a quasi-proportional-resonant controller for a multifunctional capacitive-coupling grid-connected inverter. IEEE Trans. Ind. Appl. **52**(5), 4269–4280 (2016)
9. T. Ye, N. Dai, K.W. Lao, C.-S. Lam, M.-C. Wong, Low DC voltage PV generation system with power factor correction and harmonic suppression capability in a distribution network. IET Gener. Transm. Distrib. **13**(7), 1049–1056 (2019)
10. Z. He, F. Ma, M. Li, Q. Xu, F. Zhou, Y. Yue, Multi-functional hybrid active power converter and its industrial application for electrolytic copper-foil. IET Gener. Transm. Distrib. **10**(16), 3985–3994 (2016)
11. A. Tan, K.Ç. Bayındır, M.U. Cuma, M. Tümay, Multiple harmonic elimination-based feedback controller for shunt hybrid active power filter. IET Power Electron. **10**(8), 945–956 (2017)
12. R. Bojoi, L.R. Limongi, F. Profumo, D. Roiu, A. Tenconi, Analysis of current controllers for active power filters using selective harmonic compensation schemes. IEEJ Trans. Electr. Electron. Eng. **4**(2), 139–157 (2009)
13. C. Xie, X. Zhao, K. Li, J. Zou, J.M. Guerrero, A new tuning method of multiresonant current controllers for grid-connected voltage source converters. IEEE J. Emerg. Sel. Top. Power Electron. **7**(1), 458–466 (2019)
14. C.-S. Lam, M.-C. Wong, Y.-D. Han, Hysteresis current control of hybrid active power filters. IET Power Electron. **5**(7), 1175–1187 (2012)
15. L. Wang, M.-C. Wong, C.-S. Lam, N.-Y. Dai, K.-W. Lao, C.-K. Wong, Non-linear adaptive hysteresis band pulse-width modulation control for hybrid active power filters to reduce switching loss. IET Power Electron. **8**(11), 2156–2167 (2015)
16. V.I. Utkin, *Sliding Modes in Control and Optimization* (Springer Science & Business Media, 2013)
17. IEEE Standard 519-2014, *Recommended Practices and Requirements for Harmonic Control in Electric Power Systems* (N. Y. IEEE, 2014)
18. IEEE Std 446-1995 Orange Book, *IEEE Recommended Practice for Emergency and Standby Power Systems for Industrial and Commercial Applications* (1996), pp. 1–320

Chapter 6
Multi-Quasi-Proportional-Resonant Control for TC*LC*-HAPF

6.1 Introduction

In this chapter, the modeling of the TC*LC*-HAPF is investigated based on the linear control aspect. Then, the analysis, design and implementation of a multi-quasi-proportional-resonant control (MQPRC) with gain scheduling for the TC*LC*-HAPF is proposed, which can automatically change the resonant gain between inductive and capacitive loads operation to improve its compensation performances. The proposed controller can significantly reduce the TC*LC*-HAPF steady-state current tracking errors and output current ripple. Finally, simulation and experimental results are also provided to verify the effectiveness and performance of the MQPRC for the TC*LC*-HAPF in comparison with the hysteresis current control (HCC) and quasi-proportional-resonant controller (QPRC), which shows superior compensating performances.

The contributions of this chapter are summarized in below:

(1) Investigate a fixed switching frequency controller for the TC*LC*-HAPF, which benefits of reducing output current ripple;
(2) Study the linear modeling of the TC*LC*-HAPF for the sake of the MQPRC deduction;
(3) Develop a comprehensive design criterion of the MQPRC for the TC*LC*-HAPF;

Propose a gain scheduling algorithm for MQPRC for improving the compensation performances under both inductive and capacitive loads consideration.

C.-S. Lam et al., *Advanced Controllers Design for Hybrid Active Power Filters*,
https://doi.org/10.1007/978-981-97-6847-9_6

6.2 Modeling of TC*LC*-HAPF

For the analysis, design and implementation of the MQPRC for the TC*LC*-HAPF, the modeling of the TC*LC*-HAPF is necessary. As the reactance of the coupling TC*LC* part is not a fixed value and can be varied according to the load situation, the modeling of the TC*LC* part is discussed in the following. After that, the modeling of the current control loop for the TC*LC*-HAPF is developed.

*6.2.1 Modeling of TC*LC *Part*

The main functionality of the TC*LC* part is to continuously compensate the major reactive power of the loads no matter the loads are inductive or capacitive. However, the TC*LC* part is not purely linear due to its switching component—thyristors. To simplify the controller design complexity, the model of the TC*LC* can be approximately analyzed by two linear models.

Generally, the TC*LC* part is switched between two operating modes at a fundamental period T_f, a series L_cC_{PF} filter (*LC* mode) and the L_c connected in series with an $L_{PF}//C_{PF}$ filter (*LLC* mode) to provide an equivalent reactive power with opposite polarity for cancelling out the load reactive power, that is $Q_{TCLCx} \approx Q_{cxf} = -Q_{Lxf}$. From (6.1), it can be seen that when the TC*LC* part is providing the maximum capacitive reactive power ($\alpha_x = 180°$), it works in *LC* mode. In the opposite, it works in *LLC* mode when the TC*LC* part is providing the maximum inductive reactive power ($\alpha_x = 90°$). In the middle range ($90° < \alpha_x < 180°$), the TC*LC* part will be switched between the *LC* and *LLC* modes in a proportion according to different Q_{Lxf} situation within a T_f.

$$\begin{aligned}Q_{TCLCx(\alpha_x=180°)} &= -\frac{V_x^2}{X_{C_{PF}} - X_{L_c}} \leq \\ Q_{TCLCx} &= \frac{V_{sx}^2}{X_{TCLCx}(\alpha_x)} \\ &\leq \frac{V_{sx}^2}{\frac{X_{L_{PF}}X_{C_{PF}}}{X_{C_{PF}}-X_{L_{PF}}} + X_{L_c}} \\ &= Q_{TCLCx(\alpha_x=90°)}. \end{aligned} \tag{6.1}$$

Since the switching frequency of thyristor f_T (50 Hz) is much lower than the inverter switching frequency f_{sw} (10 kHz), we can approximately treat the TC*LC* part as two linear systems (*LC* and *LLC* models) for analysis. If we can find and satisfy the stability conditions of the *LC* and *LLC* models, then the entire TC*LC*-HAPF system will be stable. Therefore, the modeling of the TC*LC* part can be separated into *LC* and *LLC* models for further analysis, in which their equivalent

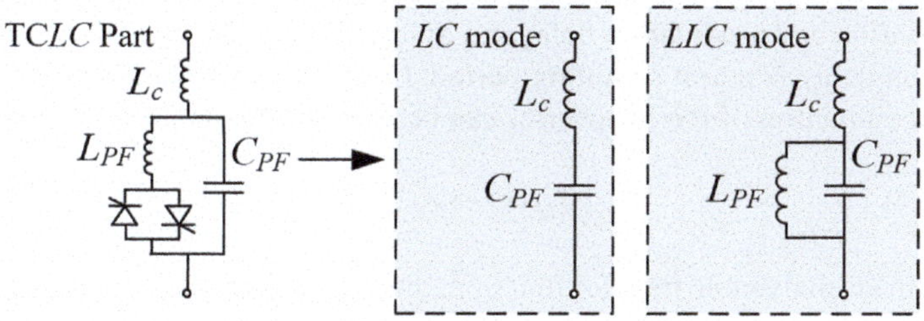

Fig. 6.1 Equivalent circuit model of the TC*LC* part

circuit model is given in Fig. 6.1. From Fig. 6.1, the Laplace domain function of the TC*LC* part $G_{TCLC}(s)$ can be developed as in (6.2).

$$G_{TCLC}(s) = \begin{cases} \frac{C_{PF}s}{L_c C_{PF}s^2+1}, & LC \text{ mode} \\ \frac{L_{PF}C_{PF}s^2+1}{L_c L_{PF}C_{PF}s^3+(L_c+L_{PF})s}, & LLC \text{ mode} \end{cases}. \tag{6.2}$$

6.2.2 Modeling of Current Control Loop

Figure 6.2 shows the current closed-loop feedback control block diagram of the TC*LC*-HAPF, where i_{cx}^* , i_{cx}, Δi_{cx} are the current reference value, the compensating current and the current tracking error, respectively. v_{refx}, v_{invx} and v_x represent the voltage reference, inverter voltage and the load voltage, respectively.

From Fig. 6.2, the current closed-loop control block diagram consists of 3 parts: current controller, PWM inverter and the TC*LC* part. The current controllers can be the proportional control (PC), QPRC, MQPRC, etc., which can offer different characteristics to the TC*LC*-HAPF. The deduction and analysis details of the MQPRC in Laplace domain will be provided in next section. In the following, the Laplace domain function of the PWM inverter will be analyzed and discussed.

In a Laplace domain model, the gain of the PWM inverter can be simplified as unity gain [1]. To precisely express the signal processing effects in the PWM inverter,

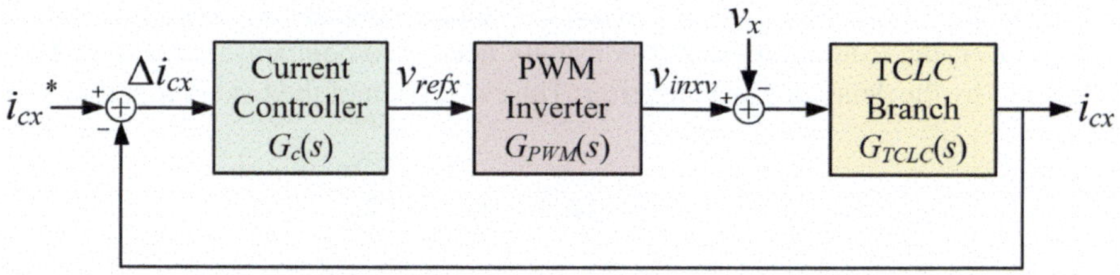

Fig. 6.2 Current closed-loop feedback control block diagram of the TC*LC*-HAPF

the computation and modulation delay in Laplace domain are considered, where the computation delay is a unit sampling period T_s and the modulation delay is $0.5T_s$. The Laplace function of PWM inverter can be expressed as [2]:

$$G_{PWM}(s) = e^{-1.5T_s \cdot s}. \tag{6.3}$$

To obtain an analyzable transfer function, the first-order Padé approximation (6.4) is applied to simplify the delay term into poles and zeros form.

$$G_{PWM}(s) = e^{-1.5T_s \cdot s} \approx \frac{1 - 0.75 \cdot T_s \cdot s}{1 + 0.75 \cdot T_s \cdot s}. \tag{6.4}$$

According to Fig. 6.2, a closed-loop transfer function of the TCLC-HAPF system can be deduced as:

$$G_{CL}(s) = \frac{G_c(s)G_{PWM}(s)G_{TCLC}(s)}{1 + G_c(s)G_{PWM}(s)G_{TCLC}(s)}. \tag{6.5}$$

The current controller transfer function $G_c(s)$ will be deduced and analyzed in the following section.

6.3 Design of Multi-Quasi-Proportional-Resonant Control for TCLC-HAPF

The MQPRC in Laplace domain is expressed as [3]:

$$G_c(s) = G_{c\_MQPR}(s) = K_p + \sum_{n=1}^{h} \frac{K_{rn}\omega_c s}{s^2 + 2\omega_c s + \omega_n^2}. \tag{6.6}$$

where K_p is the proportional gain, K_{rn} is the resonant gain at each resonant frequency, ω_c is the bandwidth of the resonant frequency, ω_n is the resonant angular frequency at each harmonic order n, and h is the highest harmonic order under consideration.

The QPRC ($h = 1$) is usually applied for improving the accuracy at the fundamental frequency, that is $\omega_n = 2\pi 50$ Hz. When the current harmonics compensation is also taken into consideration, the multiple resonant controllers should be used.

Moreover, the lead compensator is usually needed to compensate for the phase lag caused by the delay of PWM inverter. The expression of the lead compensator is [2]:

$$G_L(s) = \frac{s \cdot \cos(\phi_n) - \omega_n \cdot \sin(\phi_n)}{s}. \tag{6.7}$$

Table 6.1 Parameters of MQPRC

Symbol	Parameters
n	Harmonic order
h	Highest harmonic order under consideration
ω_n	Resonant angular frequency at n
ω_c	Bandwidth of the resonant frequency
K_p	Proportional gain
ϕ_n	Phase compensation angle
K_{rn}	Resonant gain at n

where ϕ_n is the phase compensation angle. With this phase compensation, the MQPRC transfer function in (6.6) is rewritten as (6.8).

$$G_c(s) = G_{c\_MQPR}(s) = K_p + \sum_{n=1}^{h} \frac{K_{rn}\omega_c s}{s^2 + 2\omega_c s + \omega_n^2} \cdot G_L(s). \tag{6.8}$$

From (6.8), there are total five coefficients (ω_n, ω_c, K_p, ϕ_n and K_{rn}) need to be designed for the QPRC and MQPRC and they have been listed in Table 6.1. Based on the system parameters of the TCLC-HAPF shown in Table 6.2, the coefficients' design considerations will be discussed in the following with the help of the MATLAB.

In addition, for a three-phase three-wire power system, there are not any $n = 3k$-th order harmonic currents and the common harmonic orders are $n = 6\,k \pm 1$ with $k =$

Table 6.2 System and different controllers' parameters for TCLC-HAPF in simulation and experiment

	Parameters		Value
Grid parameters	V_{sx}, f_g	Simulation	11 kV, 50 Hz
		Experiment	110 V, 50 Hz
Inverter parameters	V_{dc}, f_{sw}	Simulation	5 kV
		Experiment	60 V
TCLC parameters	L_c, L_{PF}, C_{PF}		5 mH, 30 mH, 160μF
Sampling frequency	f_s		10 kHz
Non-linear load	L_{NL}, R_{NL}		30 mH, 45 Ω
Inductive linear load	L_{LL}, R_{LL}		60 mH, 15 Ω
Capacitive linear load	C_{LL}, R_{LL}		140 μF, 20 Ω
QPRC	K_p, K_{r1}, ω_c		17.6, 1010, 6.28
MQPRC (inductive loads)	K_p, K_{r1}, K_{r5}, K_{r7}, ω_c		17.6, 850, 1220, 950, 6.28
MQPRC (capacitive loads)			17.6, 400, 1220, 950, 6.28

1, 2, …. In this section, only the 5th (250 Hz) and 7th (350 Hz) harmonic orders for the MQPRC design process are taken into consideration for simplicity.

6.3.1 Design of the Resonant Frequency ω_n and Bandwidth ω_c

According to the power grid-side fundamental frequency, harmonic frequency and its possible deviation, it is easy to design the ω_n and ω_c, which can be expressed as:

$$\omega_n = 2\pi n f_1. \tag{6.9}$$

$$\omega_c = 2\pi f_\Delta. \tag{6.10}$$

where n is harmonic order, f_1 is the power grid-side fundamental frequency and f_Δ is the margin of the frequency deviation, which can be designed according to some standard limits of frequency variation. According to the standard frequency and the limits of frequency variation of Macau and Hong Kong [4, 5], the fundamental frequency f_1 is 50 Hz and its allowable maximum frequency deviation is $\pm 2\%$, in which $f_\Delta = 1$ Hz.

6.3.2 Design of the Proportional Gain K_p

The proportional gain K_p value should be designed within the stability boundary. To find out the boundary of K_p, the closed-loop transfer function of Δi_{cx} can be derived as (6.11) based on Fig. 6.2.

$$\begin{aligned} G_e(s) &= \frac{\Delta i_{cx}(s)}{i_{cx}{}^*(s)} = \frac{1}{1 + G_c(s) G_{PWM}(s) G_{TCLC}(s)} = \frac{1}{1 + G_{c\_MQPR}(s) G_{PWM}(s) G_{TCLC}(s)} \\ &= \overbrace{\frac{1}{1 + K_p G_{PWM}(s) G_{TCLC}(s)}}^{G_{e1}(s)} \cdot \overbrace{\frac{1}{1 + \sum_{n=1}^{h} \frac{K_{rn}\omega_c s}{s^2 + 2\omega_c s + \omega_n^2} \cdot G_L(s) \cdot G_P(s) / K_p}}^{G_{e2}(s)}. \end{aligned} \tag{6.11}$$

where $G_P(s) = K_p G_{PWM}(s) G_{TCLC}(s)/[1 + K_p G_{PWM}(s) G_{TCLC}(s)]$. $G_P(s)$ can be considered as the closed-loop transfer function of the proportional control (PC) for the TCLC-HAPF. From (6.11), the system is stable when the following two conditions are satisfied.

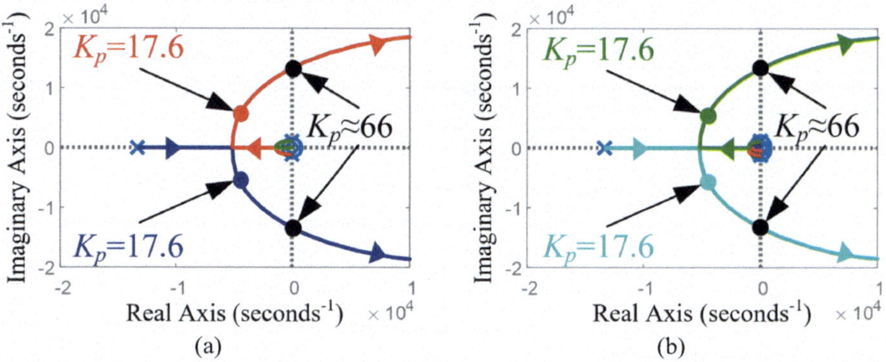

Fig. 6.3 Root locus of $G_{e1}(s)$ under two modes for MQPRC: **a** *LC* mode; **b** *LLC* mode

(1) the roots of $1 + K_pG_{PWM}(s)G_{TCLC}(s) = 0$ are located at the left-hand side of the s-plane;
(2) the roots of $1 + \sum_{n=1}^{h} \frac{K_{rn}\omega_c s}{s^2+2\omega_c s+\omega_n^2} \cdot G_L(s) \cdot G_P(s) \big/ K_p = 0$ are also located at the left-hand side of the s-plane.

The first stability criterion can be used to find out the boundary value of K_p, while the second criterion can be used to confirm the boundary value of K_{rn}. By applying the Routh's stability criterion on $G_{e1}(s)$, the boundary of K_p can be deduced as (6.12) for both *LC* and *LLC* models.

$$0 < K_P < \frac{L_c C_{PF} - (0.75T_s)^2}{0.75T_s C_{PF}}. \qquad (6.12)$$

The root locus of $K_pG_{PWM}(s)G_{TCLC}(s)$ is depicted in Fig. 6.3. The upper boundary of K_p is around 66 for both *LC* and *LLC* models from the simulation, which also satisfies the inequality of (6.12). Nevertheless, K_p is selected to be 17.6 to obtain sufficient damping ($\zeta = 0.707$) for avoiding the system oscillation [6]. The bode plots under different K_p of the closed-loop transfer function $G_P(s)$ are presented in Fig. 6.4. It can be observed that the system bandwidth increases along with the K_p increases. Once the K_p continues to increase, a resonant peak will appear at the cut-off frequency. Therefore, there is a trade-off between the system bandwidth and high frequency rejection. Based on the above studies, $K_p = 17.6$ is chosen for getting a balance between the system bandwidth and high frequency rejection.

6.3.3 Design of the Phase Compensation Angle ϕ_n

The phase compensation angle ϕ_n is used to compensate for the phase lag phenomenon caused by the delay of PWM inverter. It can be set to the phase lag of the open-loop transfer function of $G_P(s)$ at each of the harmonic resonant frequency

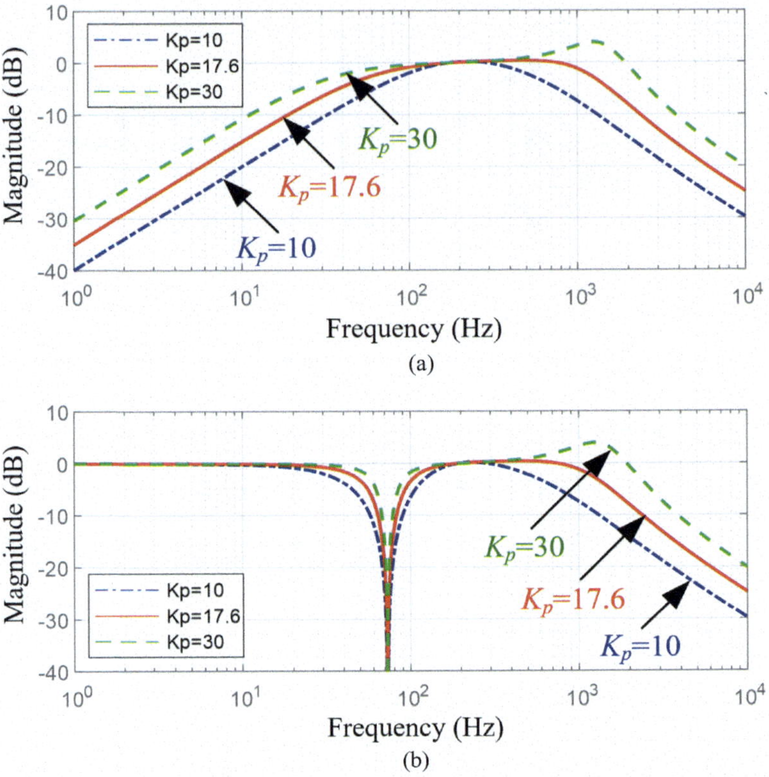

Fig. 6.4 Closed-loop bode diagram of $G_P(s)$ under different K_p values: **a** *LC* mode; **b** *LLC* mode

to guarantee the zero-phase error. Because the ϕ_n is proportional to the resonant frequency and system delay [7], it is set to $1.5T_s\omega_n$ in this book since the time delay of the system is $1.5T_s$ as discussed in Sect. 6.3.2. Then, (6.7) can be written as:

$$G_L(s) = \frac{s \cdot \cos(1.5T_s\omega_n) - \omega_n \cdot \sin(1.5T_s\omega_n)}{s}. \tag{6.13}$$

6.3.4 *Design of the Resonant Gain* $\mathbf{K_{rn}}$

Once the K_p and ϕ_n are designed, the K_{rn} can be determined accordingly. The value of the resonant gain K_{rn} should be designed to satisfy the transient response at the resonant frequency and within the boundary. According to the second stability criterion of (6.11), the boundary value of K_{rn} for the MQPRC can be obtained. The root locus diagram of $\sum_{n=1}^{h} \frac{K_{rn}\omega_c s}{s^2+2\omega_c s+\omega_n^2} \cdot G_L(s) \cdot G_P(s) \big/ K_p$ with the K_p and ϕ_n values

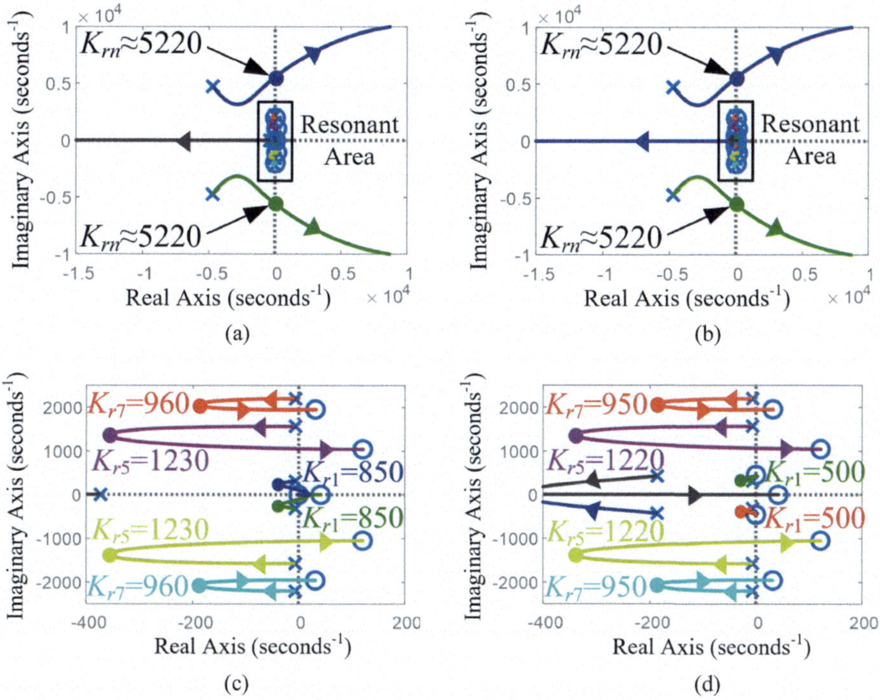

Fig. 6.5 Root locus of $G_{e2}(s)$ under two modes for MQPRC: **a** *LC* mode; **b** *LLC* mode; **c** zoomed-in resonant area of *LC* mode; **d** zoomed-in resonant area of *LLC* mode

designed in Sects. 6.3.2 and 6.3.3 are shown in Fig. 6.5a–b, which can be observed the boundary of K_{rn} in the MQPRC under two different models. From the simulation results and Fig. 6.5a–b, the boundary of K_{rn} under two models of the MQPRC are the same, which is about 5220.

However, the large resonant gain does not imply good system response [6]. The selection of the K_{rn} is based on transient response speed which is determined by the system decay rate. The larger the decay rate is, the faster response the system obtains [8]. Generally, the decay rate is the real part of each pole of the closed-loop transfer function. To find out the corresponding K_{rn}, Fig. 6.5c–d show the zoomed-in resonant area of the corresponding root locus diagram. The shape of each locus of the resonant part is a convex curve and it has a unique peak located at the left-hand side, which is the point that has the largest decay rate. The gain at that point can provide the fastest response speed and has the largest stability margin for the system. Based on this consideration, the resonant gain can be obtained from Fig. 6.5c–d. The K_{r1}, K_{r5} and K_{r7} under *LC* model for the MQPRC are 850, 1230 and 960, respectively, which can be found in Fig. 6.5c. For the *LLC* model as shown in Fig. 6.5d, the K_{r1}, K_{r5} and K_{r7} are 400, 1220 and 950, respectively. Since there are two different K_{r1} for two different operating modes, the controller needs to select the gain automatically according to the load situation, and the corresponding control strategy will be presented in the next section.

At the fundamental frequency, the resonant gains for both *LC* and *LLC* modes are very different. On the other hand, at the harmonic frequency, the boundary of the resonant gains and the resonant gains for both *LC* and *LLC* modes are approximately the same, which means that we can design the resonant gains at the harmonic frequency in either *LC* or *LLC* operating mode.

6.4 Overall Control Block Diagram

According to the previous works [9–11], the TC*LC* part and the inverter part should be controlled coordinately. The overall control strategy of the TC*LC*-HAPF is summarized in Fig. 6.6, which can be separated into two parts: TC*LC* part control and active inverter part control.

The TC*LC* part can compensate for most of the load reactive power no matter the load is inductive or capacitive by adaptively changing the equivalent fundamental reactance X_{TCLCx}. The X_{TCLCx} is changed by controlling the final firing angle $\alpha_x \in [90°, 180°]$ of two thyristors in each phase, where the α_x is determined according to the initial firing angle α_{0x} and the phase angle of the voltage across TC*LC* part ϕ_x. The relationship between the X_{TCLCx} and α_{0x} is given as (6.14). Besides, the relationship between the X_{TCLCx} and the load reactive power Q_{Lx} is already shown in (1.6).

$$X_{TCLCx}(\alpha_{0x}) = \frac{\pi X_{L_{PF}} X_{C_{PF}}}{X_{C_{PF}}(2\pi - 2\alpha_{0x} + \sin 2\alpha_{0x}) - \pi X_{L_{PF}}} + X_{L_c}. \tag{6.14}$$

The Q_{Lx} can be obtained by the following equations [12], where $\overline{q}_{Lx}$ is the low-passed value of q_{Lx}, v_{xf} is the fundamental load voltage after passing through the band-pass filter.

$$\begin{bmatrix} q_{La} \\ q_{Lb} \\ q_{Lc} \end{bmatrix} = \begin{bmatrix} v_{bf} \cdot i_{Lc} - v_{cf} \cdot i_{Lb} \\ v_{cf} \cdot i_{La} - v_{af} \cdot i_{Lc} \\ v_{af} \cdot i_{Lb} - v_{bf} \cdot i_{La} \end{bmatrix}. \tag{6.15}$$

$$Q_{Lx} \approx -\frac{\overline{q}_{Lx}}{2}. \tag{6.16}$$

A lookup table (LUT) has been installed inside the controller because (6.14) does not have a closed-form solution. Once the equivalent fundamental reactance $X_{TCLCx}(\alpha_{0x})$ is obtained from (1.6), the initial firing angle α_{0x} can be extracted based on the LUT of (6.14). As the control of the firing angles of the adaptive filter part is usually referenced to the phase angle of the load voltage v_x of each phase, an angle transformation process is required for getting the final firing angle a_x [12]. The a_x can

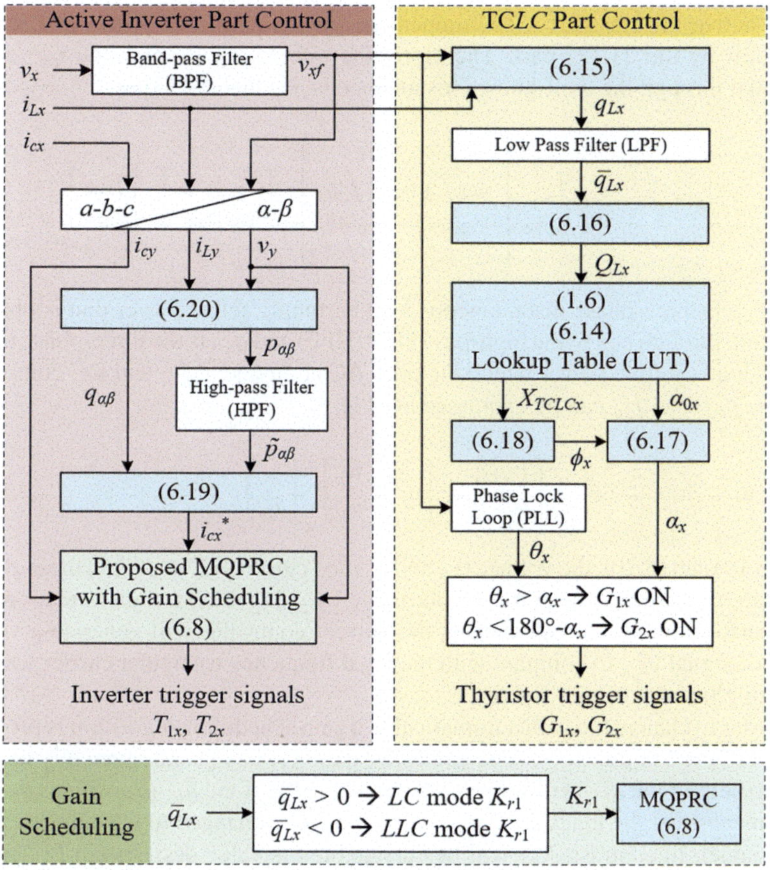

Fig. 6.6 Overall control block diagram of TCLC-HAPF for MQPRC with gain scheduling

be obtained as (6.17), where ϕ_x is the phase angle of the voltage across the adaptive filter presented in (6.18). Finally, the α_x is then compared with the load voltage phase angle θ_x to generate the thyristor trigger signals G_{1x} and G_{2x} of each phase.

$$\begin{bmatrix} \alpha_a \\ \alpha_b \\ \alpha_c \end{bmatrix} = \begin{bmatrix} \alpha_{0a} \\ \alpha_{0b} \\ \alpha_{0c} \end{bmatrix} - \begin{bmatrix} \phi_a \\ \phi_b \\ \phi_c \end{bmatrix}. \tag{6.17}$$

$$\begin{bmatrix} \phi_a \\ \phi_b \\ \phi_c \end{bmatrix} = \begin{bmatrix} \tan^{-1}\left(\frac{X_{TCLCc}-X_{TCLCb}}{\sqrt{3}(X_{TCLCc}+X_{TCLCb})}\right) \\ \tan^{-1}\left(\frac{X_{TCLCa}-X_{TCLCc}}{\sqrt{3}(X_{TCLCa}+X_{TCLCc})}\right) \\ \tan^{-1}\left(\frac{X_{TCLCb}-X_{TCLCa}}{\sqrt{3}(X_{TCLCb}+X_{TCLCa})}\right) \end{bmatrix}_{\tan^{-1}\theta\in[-90^\circ,90^\circ]}. \tag{6.18}$$

The active inverter part is to compensate for the harmonic power and reactive power left by the TC*LC* part. Therefore, the reference current value $i_{cx}{}^{*}$ can be calculated through the well-known instantaneous *p*-*q* theory [13] as:

$$\begin{bmatrix} i_{ca}{}^{*} \\ i_{cb}{}^{*} \\ i_{cc}{}^{*} \end{bmatrix} = \sqrt{\frac{2}{3}} \cdot \frac{1}{v_{\alpha}^{2} + v_{\beta}^{2}} \begin{bmatrix} 1 & 0 \\ -1/2 & \sqrt{3}/2 \\ -1/2 & -\sqrt{3}/2 \end{bmatrix} \cdot \begin{bmatrix} v_{\alpha} & -v_{\beta} \\ v_{\beta} & v_{\alpha} \end{bmatrix} \cdot \begin{bmatrix} \tilde{p}_{\alpha\beta} \\ q_{\alpha\beta} \end{bmatrix}. \tag{6.19}$$

where $\tilde{p}_{\alpha\beta}$ is three-phase instantaneous load harmonic active power that is obtained by passing through $p_{\alpha\beta}$ to the high-pass filter (HPF) and $q_{\alpha\beta}$ is the three-phase instantaneous load reactive power which includes AC component $\tilde{q}_{\alpha\beta}$ and DC component $\overline{q}_{\alpha\beta}$. The $p_{\alpha\beta}$ and $q_{\alpha\beta}$ can be expressed as [11]:

$$\begin{bmatrix} p_{\alpha\beta} \\ q_{\alpha\beta} \end{bmatrix} = \begin{bmatrix} v_{\alpha} & v_{\beta} \\ -v_{\beta} & v_{\alpha} \end{bmatrix} \cdot \begin{bmatrix} i_{\alpha} \\ i_{\beta} \end{bmatrix}. \tag{6.20}$$

In (6.19) and (6.20), the voltage (v_{α} and v_{β}) and currents (i_{α} and i_{β}) in the α-β coordinate are transformed from *a*-*b*-*c* coordinate by using Clarke transformation [13]. After performing all the calculation, the current controller will generate a voltage reference signal v_{refx} to compare with the fixed frequency triangular carrier wave for generating inverter trigger signals.

In order to change the gain automatically, a gain scheduling algorithm is proposed for the MQPRC in this book. The controller gain K_{r1} can be automatically changed with the help of the $\overline{q}_{Lx}$ getting from (6.16) with LPF. If the $\overline{q}_{Lx}$ is positive ($\overline{q}_{Lx} > 0$), which means that the loads are inductive, K_{r1} will be changed to the value designed in *LC* model. In contrast, K_{r1} will be changed to the value designed in *LLC* model when the $\overline{q}_{Lx}$ is negative ($\overline{q}_{Lx} < 0$). Then the MQPRC will work with different K_{r1} according to the load situation.

6.5 Simulation Results

In this section, simulation results of applying different controllers for the TC*LC*-HAPF under 11 kV system voltage are provided by using PSCAD/EMTDC. Table 6.2 shows the system and different controllers' parameters for simulation, in which the DC-link voltage is modelled by a DC voltage source to simplify the verification of the validity and effectiveness of the MQPRC with proposed gain scheduling algorithm for the TC*LC*-HAPF. The simulation results are presented in Figs. 6.7, 6.8 and 6.9 and summarized in Table 6.3. All the settling time T_{st} of this work is obtained after turning on the TC*LC*-HAPF to the system with loads already connected.

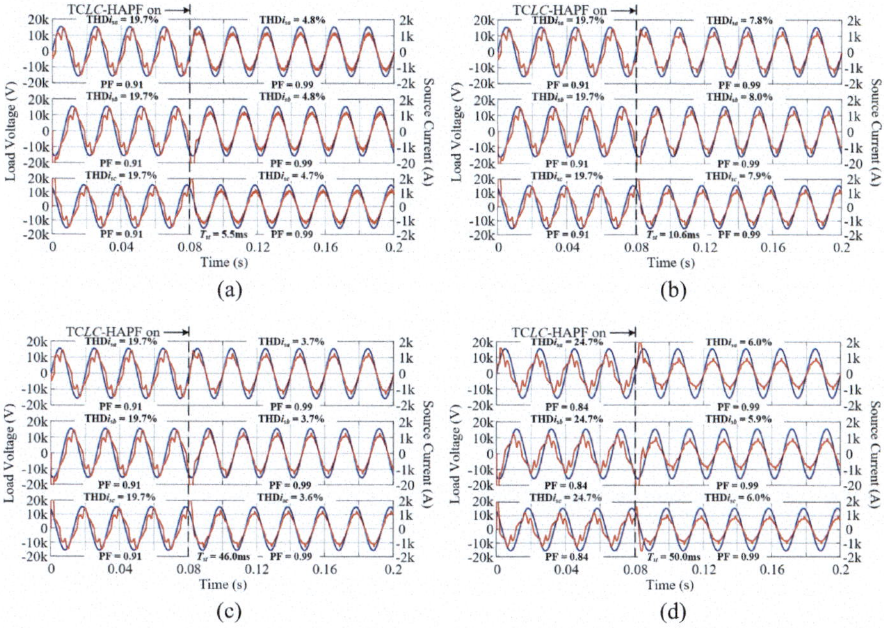

Fig. 6.7 Simulation results of TC*LC*-HAPF dynamic performance by using different controller: **a** HCC under inductive loads; **b** QPRC under inductive loads; **c** MQPRC with proposed gain scheduling under inductive loads; **d** MQPRC with proposed gain scheduling under capacitive loads

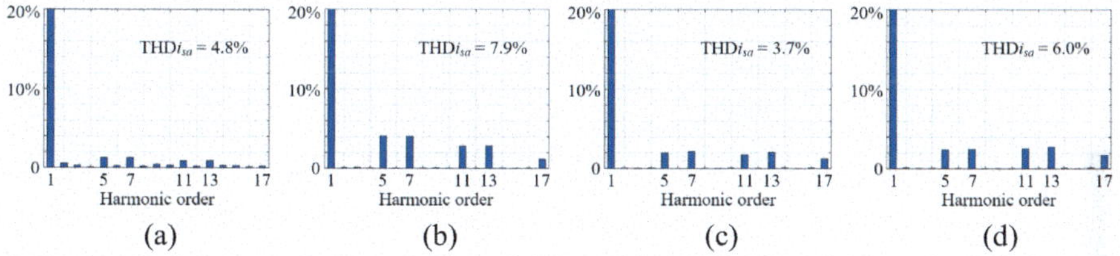

Fig. 6.8 Simulation results of i_{sa} harmonic spectrum after compensation. **a** HCC under inductive loads; **b** QPRC under inductive loads; **c** MQPRC with proposed gain scheduling under inductive loads; **d** MQPRC with proposed gain scheduling under capacitive loads

(A) ***Hysteresis Current Control***: Figs. 6.7a, 6.8a and Table 6.3 show the performance of the HCC under inductive loads compensation. Once the TC*LC*-HAPF starts operation, it provides a very fast transient response. The compensated THDi_{sx} is about 4.8%. And the significant current ripple emerges due to the variable switching frequency characteristic.

(B) ***Quasi-Proportional-Resonant Control***: Figs. 6.7b, 6.8b and Table 6.3 show the performance of the QPRC under inductive loads compensation. As the resonant peak of the QPRC is tuned at the fundamental frequency only, the compensated THDi_{sx} is about 8%, which is the worst among the controllers.

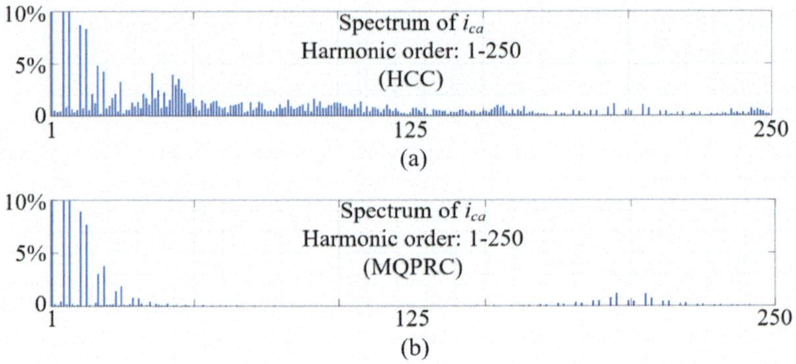

Fig. 6.9 Simulation results of phase *a* compensating current i_{ca} harmonics spectrum: **a** HCC and **b** MQPRC

Table 6.3 Simulation results before and after TC*LC*-HAPF compensation

		P_{sx} (kW)	Q_{sx} (kvar)	PF	i_{sx} (kA)	Worst THDi_{sx} (%)	T_{st} (ms)
Inductive Loads	Before Compensation	7904.9	3137.2	0.91	0.88	19.7	/
	HCC	7887.8	15.5	0.99	0.80	4.8	5.5
	QPRC	7858.7	12.0	0.99	0.80	8.0	10.6
	MQPRC	7832.4	9.7	0.99	0.79	3.7	46.0
Capacitive Loads	Before Compensation	5764.3	-3716.8	0.84	0.71	24.7	/
	HCC	5895.6	-16.2	0.99	0.60	7.3	5.8
	QPRC	5880.5	-11.6	0.99	0.60	10.7	10.8
	MQPRC	5865.1	-10.8	0.99	0.59	6.0	50.0

(C) ***Multi-Quasi-Proportional-Resonant Control***: Figs. 6.7c, 6.8c and Table 6.3 show the performance of the MQPRC with proposed gain scheduling algorithm under inductive loads compensation. Once the TC*LC*-HAPF starts operation, the response is the slowest among the controllers because of the integrators. However, the steady-state compensation performance is much better than that of the QPRC and HCC as it consists of resonant peaks at considered harmonic frequency orders. Hence, the THDi_{sx} can be reduced to about 3.7%, which is the best among the existing control strategy of TC*LC*-HAPF.

Figure 6.7d, 6.8d and Table 6.3 show the performance of the MQPRC with proposed gain scheduling algorithm under capacitive loads. The THDi_{sx} after compensation is about 6%, it is a bit higher than that of inductive loads because of the fundamental current reduction. It tells the effectiveness of the MQPRC with proposed gain scheduling algorithm for TC*LC*-HAPF.

Table 6.3 also shows the simulation results of different controllers under capacitive loads, which clearly shows that the MQPRC still performs the best compensation ability under capacitive loads by using the proposed gain scheduling algorithm.

Figure 6.9 shows the harmonics spectrum of the phase a compensating current, in which the fixed switching frequency MQPRC methods can obtain lower switching noise than the hysteresis current control method.

6.6 Experimental Results

Due to the limitation of the laboratory facilities, a scale down small rating three-phase three-wire 110V-5kVA TC*LC*-HAPF experimental prototype was built in the laboratory and shown in Fig. 6.10. Table 6.2 shows the experimental setting of the system and controllers' parameters. The digital control system of the TC*LC*-HAPF consists of two digital signal processors (DSPs) TMS320F2812s for active inverter part and TC*LC* part control. The switching device employed in TC*LC*-HAPF are Mitsubishi IGBT intelligent power modules PM300DSA60 for active inverter part and SanRex PK110FG160 thyristors for TC*LC* part, respectively. The basic settings of both DSPs are the same and the sampling frequency is set as 10 kHz. For every 1/10 kHz (s) period, the timer will provide an interrupt signal to trigger the analog-to-digital (A/D) conversion, in which, the load voltage, load current and compensating current signals from the sensing circuits are digitized and used for computing the control algorithms, such as instantaneous *p*-*q* theory, HCC, QPRC or MQPRC with gain scheduling, TC*LC* part control, etc. Finally, the corresponding switching signals will be generated for the IGBT and thyristor modules. The experimental results are presented in Figs. 6.11, 6.12 and 6.14 and summarized in Table 6.4.

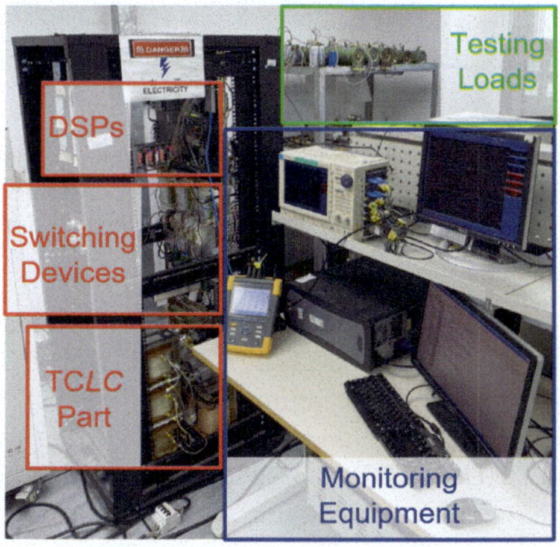

Fig. 6.10 Experimental setup of an 110V–5kVA TC*LC*-HAPF experimental prototype

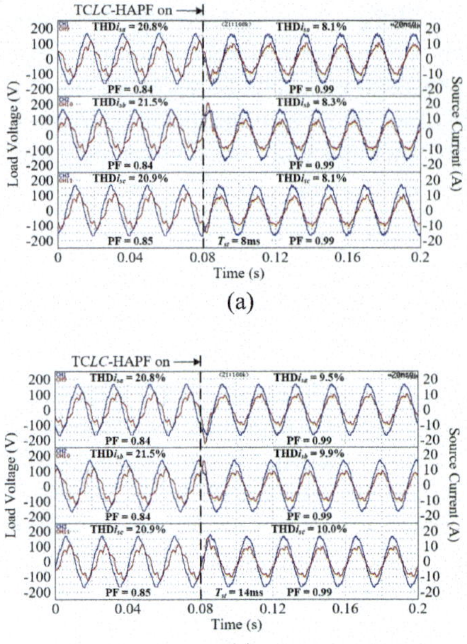

Fig. 6.11 Experimental results of TC*LC*-HAPF dynamic performance by using different controller: **a** HCC under inductive loads; **b** QPRC under inductive loads; **c** MQPRC with proposed gain scheduling under inductive loads; **d** MQPRC with proposed gain scheduling control under capacitive loads

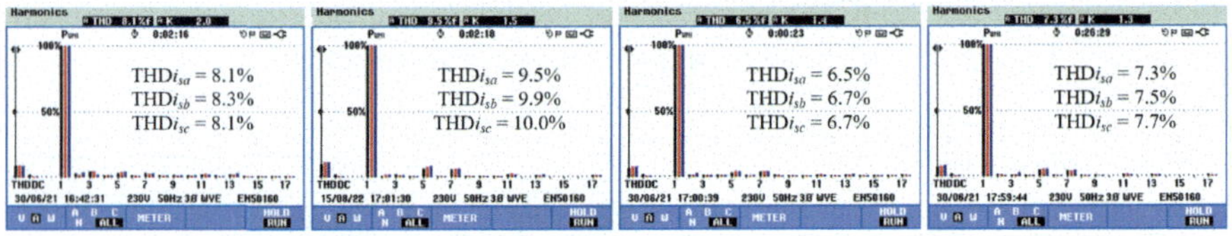

Fig. 6.12 Experimental results of i_{sx} harmonic spectrum after compensation. **a** HCC under inductive loads; **b** QPRC under inductive loads; **c** MQPRC with proposed gain scheduling under inductive loads; **d** MQPRC with proposed gain scheduling control under capacitive loads

In addition, regarding the IEEE standard 519-2014 [14], the acceptable total demand distortion (TDD) $\leq 15\%$ with I_{sc}/I_L is 100 < 1000 scale (a small experimental prototype rating of 110V–5kVA). The nominal rate current is assumed to be equal to the fundamental load current for the worst-case analysis, which results in THD = TDD $\leq 15\%$. Therefore, this book evaluates the TC*LC*-HAPF current harmonics compensating performance by setting acceptable THD $\leq 15\%$.

(A) ***Hysteresis Current Control***: Figs. 6.11a, 6.12a and Table 6.4 show the performance of the HCC under inductive loads compensation. Once the TC*LC*-HAPF

Table 6.4 Experimental results before and after TC*LC*-HAPF compensation

		P_{sx} (W)	Q_{sx} (var)	PF	i_{sx} (A)	Worst THDi_{sx} (%)	T_{st} (ms)
Inductive Loads	Before Compensation	680	400	0.84	7.3	21.5	/
	HCC	730	20	0.99	6.5	8.3	8
	QPRC	730	10	0.99	6.5	10.0	14
	MQPRC	720	10	0.99	6.4	6.7	70
Capacitive Loads	Before Compensation	770	−240	0.93	7.1	22.8	/
	HCC	800	−20	0.99	7.0	9.2	8
	QPRC	800	−10	0.99	7.1	9.9	12
	MQPRC	800	−10	0.99	6.9	7.7	76

starts operation, it provides a very fast transient response. The compensated THDi_{sx} is about 8.1%, and the significant current ripple emerges.

(B) ***Quasi-Proportional-Resonant Control***: Figs. 6.11b, 6.12b and Table 6.4 show the performance of the QPRC under inductive loads compensation. The harmonic frequencies are not compensated, which leads the compensated THDi_{sx} to be about 10.0%, which is the worst among the controllers.

(C) ***Multi-Quasi-Proportional-Resonant Control***: Figs. 6.11c, 6.12c and Table 6.4 show the performance of the MQPRC with proposed gain scheduling algorithm under inductive loads compensation. Once the TC*LC*-HAPF starts operation, the response is the slowest because of the integrators. However, the steady-state compensation performance is the best among the controllers as it consists of resonant peaks at considered harmonic frequency orders. Hence, the THDi_{sx} can be reduced to about 6.7%, which is the best among the existing control strategy of TC*LC*-HAPF.

Figures. 6.11d, 6.12d and Table 6.4 show the performance of the MQPRC under capacitive loads. The settling time is similar to that of the inductive loads case and the compensated THDi_{sx} can be reduced to 7.5% from 22%, which shows the effectiveness of the MQPRC with proposed gain scheduling algorithm.

If the resonant gain K_{r1} of the MQPRC for the inductive load ($K_{r1} = 850$) compensation is also applied for the capacitive loads compensation, that means without the proposed gain scheduling algorithm ($K_{r1} = 400$), this K_{r1} gain value will deteriorate the TC*LC*-HAPF compensation performance (worst THDi_{sx} = 10.9%), as verified by Fig. 6.13, which shows the effectiveness of the gain scheduling algorithm.

Table 6.4 also shows the experimental results with different controllers under capacitive loads, which clearly illustrates that the MQPRC still performs the best compensation results under capacitive loads by using the proposed gain scheduling algorithm.

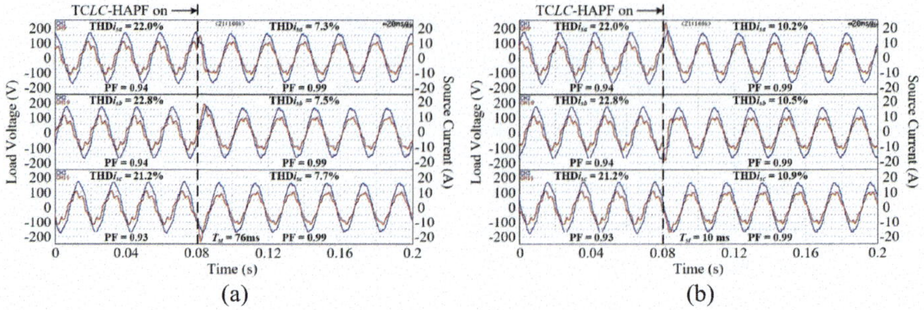

Fig. 6.13 Experimental results of TC*LC*-HAPF steady-state performance by using MQPRC: **a** without proposed gain scheduling under capacitive loads; **b** with proposed gain scheduling under capacitive loads

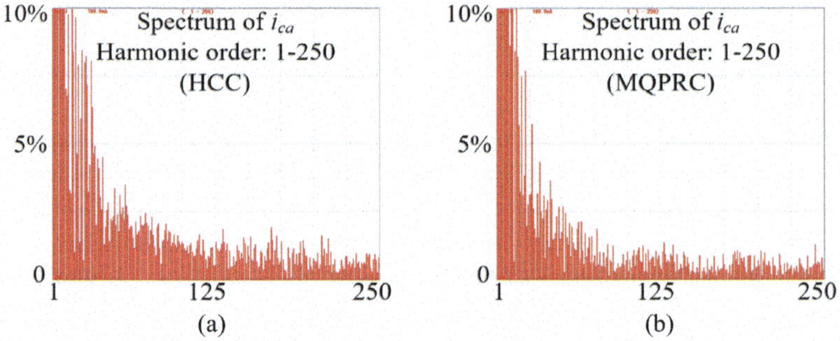

Fig. 6.14 Experimental results of phase a compensating current i_{ca} harmonics spectrum: **a** HCC and **b** MQPRC

Figure 6.14 illustrates that the HCC obtains larger output current ripple and noise, which is due to its wide and variable switching frequency characteristics, while the MQPRC can obtain lower switching noise due to its fixed switching frequency.

6.7 Summary

This chapter proposes the analysis, design and implementation of the MQPRC with gain scheduling algorithm for the TC*LC*-HAPF, which can improve the steady-state performance with lower steady-state error, lower output current ripple and noise under both inductive and capacitive loads, in comparison with the HCC and QPRC. First, the mathematical model and closed-loop control block for the TC*LC*-HAPF with MQPRC are derived. Then, the parameter design criteria of the MQPRC for the TC*LC*-HAPF are presented and discussed. After that, the gain scheduling algorithm is proposed for the MQPRC of the TC*LC*-HAPF, which can improve the compensation performance under both inductive and capacitive loads consideration. Simulations and experiments are conducted to verify the effectiveness and performances of the MQPRC with proposed gain scheduling algorithm.

References

1. T. Ye, N. Dai, C.-S. Lam, M.-C. Wong, J.M. Guerrero, Analysis, design, and implementation of a quasi-proportional-resonant controller for a multifunctional capacitive-coupling grid-connected inverter. IEEE Trans. Ind. Appl. **52**(5), 4269–4280 (2016)
2. S. Buso, P. Mattavelli, *Digital Control in Power Electronics* (Morgan & Claypool, 2015)
3. L. Herman, I. Papic, B. Blazic, A proportional-resonant current controller for selective harmonic compensation in a hybrid active power filter. IEEE Trans. Power Del. **29**(5), 2055–2065 (2014)
4. CEM—Companhia de Electricidade de Macau, "Standard Conditions of Supply. https://www.cem-macau.com/en/customer-service/energy-supply/standard-condition-of-supply
5. CPL Power Hong Kong Ltd., Supply Rules. https://www.clp.com.hk/en/customer-service/open-and-closeaccount/supply-rules
6. C. Xie, X. Zhao, K. Li, J. Zou, J.M. Guerrero, A new tuning method of multiresonant current controllers for grid-connected voltage source converters. IEEE Trans. Emerg. Sel. Top. Power Electron. **7**(1), 458–466 (2019)
7. R.I. Bojoi, G. Griva, V. Bostan, M. Guerriero, F. Farina, F. Profumo, Current control strategy for power conditioners using sinusoidal signal integrators in synchronous reference frame. IEEE Trans. Ind. Electron. **20**(6), 1402–1412 (2005)
8. R.C. Dorf, R.H. Bishop, *Modern Control Systems*, 12th edn. (Pearson Education, 2010)
9. L. Wang, C.-S. Lam, M.-C. Wong, A hybrid-STATCOM with wide compensation range and low DC-Link voltage. IEEE Trans. Ind. Electron. **63**(6), 3333–3343 (2016)
10. L. Wang, C.-S. Lam, M.-C. Wong, Modeling and parameter design of thyristor-controlled LC-coupled hybrid active power filter (TCLC-HAPF) for unbalanced compensation. IEEE Trans. Ind. Electron. **64**(3), 1827–1840 (2017)
11. L. Wang, C.-S. Lam, M.-C. Wong, Selective compensation of distortion, unbalanced and reactive power of a thyristor-controlled LC—coupling hybrid active power filter (TCLC-HAPF). IEEE Trans. Power Electron. **32**(12), 9065–9077 (2017)
12. L. Wang, C.-S. Lam, M.-C. Wong, Unbalanced control strategy for a thyristor-controlled LC-coupling hybrid active power filter in three-phase three-wire systems. IEEE Trans. Power Electron. **32**(2), 1056–1069 (2017)
13. H. Akagi, Y. Kanazawa, A. Nabae, Instantaneous reactive power compensators comprising switching devices without energy storage components. IEEE Trans. Ind. Appl. **IA-20**(3), 625–630 (1984)
14. IEEE Standard 519-2014, *IEEE Recommended Practices and Requirements for Harmonic Control in Electrical Power Systems* (2014)

Chapter 7
H_∞ Control for TC*LC*-HAPF

7.1 Introduction

In this chapter, a harmonic state-space (HSS) modeling is firstly proposed to address the nonlinearity issue of the TC*LC*-HAPF. The HSS based TC*LC*-HAPF model is linear time-invariant (LTI), which enables the optimal control techniques. Then, the linear matrix inequality (LMI) based H_∞ optimal control (HOC) design is proposed, which finds the optimal controller parameters via convex optimization algorithm. The proposed controller for the TC*LC*-HAPF achieves high steady-state accuracy, fast transient response, and fixed switching frequency simultaneously. Finally, simulation and experimental results are also provided to verify the effectiveness and performance of the optimal controller for the TC*LC*-HAPF in comparison with the recent developed state-of-the-art multi-quasi-proportional-resonant controller (MQPRC), which shows superior control performances.

The contribution of this chapter is summarized as follows:

(1) Propose a HSS model for the TC*LC*-HAPF, in which the nonlinear TC*LC*-HAPF can be modeled as a linear time-invariant (LTI) system, thus enables its optimal control design methods.
(2) Propose an LMI based HOC design for the TC*LC*-HAPF with low steady-state error, fast transient response, and fixed switching frequency.
(3) Propose an overall control block diagram for the TC*LC*-HAPF with HOC.
(4) Verify the performance and robustness of the proposed HOC for the TC*LC*-HAPF by simulation and experimental results in comparison with the state-of-the art MQPRC.

C.-S. Lam et al., *Advanced Controllers Design for Hybrid Active Power Filters*,
https://doi.org/10.1007/978-981-97-6847-9_7

7.2 System Modeling and Harmonic State Space Modeling of TCLC-HAPF

From the circuit configuration of the three-phase TC*LC*-HAPF as shown in Fig. 1.5, the time-domain system model of the TC*LC*-HAPF can be derived as follows [1], where i_{gx} represents the current flowing through the thyristor branch.

$$L_c \frac{di_{cx}}{dt} = v_x - v_{cx} - v_{invx} \tag{7.1}$$

$$C_{PF} \frac{dv_{cx}}{dt} = i_{cx} - i_{gx} \tag{7.2}$$

$$L_{PF} \frac{di_{gx}}{dt} = G_x(t) v_{cx} \tag{7.3}$$

Based on the TC*LC*-HAPF system model (7.1)–(7.3), the time-domain state-space model of TC*LC*-HAPF is provided as [2]:

$$\frac{dx_x}{dt} = A_x(t) x_x + B_{ux} u_x + B_{wx} w_x \tag{7.4}$$

$$y_x = C_x x_x + D_x w_x \tag{7.5}$$

where $x_x = [i_{cx}, v_{cx}, i_{gx}]^T$ are the vector of state variables, $u_x = v_{invx}$ is considered as the control input, $w_x = [i_{cx}^{*}, v_x]^T$ contains the reference compensating currents i_{cx}^{*} and load voltages, y_x is defined as the compensating current error signal $\Delta i_{cx} = i_{cx}^{*} - i_{cx}$, and the time-domain state-space matrices A_x, B_{ux}, B_{wx}, C, and D are given in the follows:

$$\begin{cases} A_x(t) = \begin{bmatrix} 0 & -1/L_c & 0 \\ 1/C_{PF} & 0 & -1/C_{PF} \\ 0 & G_x(t)/L_{PF} & 0 \end{bmatrix} \\ B_{ux} = \begin{bmatrix} -1/L_c \\ 0 \\ 0 \end{bmatrix}, B_{wx} = \begin{bmatrix} -1/L_c & 0 \\ 0 & 0 \\ 0 & 0 \end{bmatrix} \\ C_x = \begin{bmatrix} 1\ 0\ 0 \end{bmatrix}, D_x = \begin{bmatrix} 0\ -1 \end{bmatrix} \end{cases} \tag{7.6}$$

As the state-space matrix A_x contains the thyristor on–off state signal G_x, the system described by (7.4)–(7.6) is not an LTI system. To address this concern, the HSS model [3–5] of the TC*LC*-HAPF is proposed in the following.

The HSS modeling technique is based on the complex Fourier series, where an arbitrary signal $x(t)$ with the fundamental angular frequency ω_1 is given by the series:

$$x(t) = \sum_{n=-\infty}^{+\infty} X_n e^{jn\omega_1 t} \tag{7.7}$$

where X_n is the n-th order harmonic coefficient of the series, which can be calculated by:

$$X_n = \frac{1}{T}\int_{t_0}^{t_0+T_f} x(t)e^{-jn\omega_1 t}dt \tag{7.8}$$

where T_f is the fundamental period.

Based on the Fourier series and harmonic balance theory [6], the time-domain state-space TC*LC*-HAPF model (7.4)–(7.6) can be transformed into the frequency-domain HSS model, as shown in Fig. 7.1 that can be derived as [7–12]:

$$\frac{dX_x}{dt} = A_x^H X_x + B_{ux}^H U_x + B_{wx}^H W_x \tag{7.9}$$

$$Y_x = C_x^H X_x + D_x^H W_x \tag{7.10}$$

where X_x, U_x, and W_x are the vectors of harmonic coefficients of the corresponding time-domain signals x_x, u_x, and w_x, which are given as:

$$\begin{aligned} X_x &= [\ldots\ X_{x-2}\ X_{x-1}\ X_{x0}\ X_{x1}\ X_{x2}\ \ldots]^T \\ U_x &= [\ldots\ U_{x-2}\ U_{x-1}\ U_{x0}\ U_{x1}\ U_{x2}\ \ldots]^T \\ W_x &= [\ldots\ W_{x-2}\ W_{x-1}\ W_{x0}\ W_{x1}\ W_{x2}\ \ldots]^T \end{aligned} \tag{7.11}$$

And the HSS matrices are given as follows:

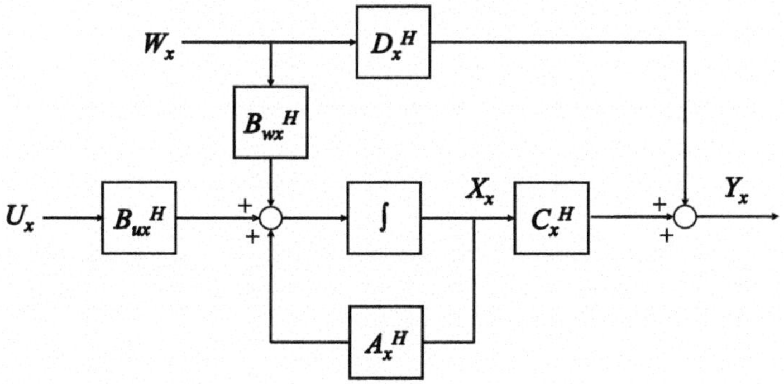

Fig. 7.1 Block diagram of TC*LC*-HAPF HSS model

$$
\begin{aligned}
N &= diag\left[..., -j2\omega_1, -j\omega_1, 0, j\omega_1, j2\omega_1, ...\right] \\
G_x^H &= \begin{bmatrix} \ddots & \vdots & \vdots & \vdots & ⋰ \\ \cdots & G_{x0} & G_{x-1} & G_{x-2} & \cdots \\ \cdots & G_{x1} & G_{x0} & G_{x-1} & \cdots \\ \cdots & G_{x2} & G_{x1} & G_{x0} & \cdots \\ ⋰ & \vdots & \vdots & \vdots & \ddots \end{bmatrix} \\
A_x^H &= \begin{bmatrix} -N & -1/L_F \cdot I & O \\ -1/C_F \cdot I & -N & -1/C_F \cdot I \\ O & T_x^H/L_T & -N \end{bmatrix} \\
B_{ux}^H &= \begin{bmatrix} -1/L_F \cdot I \\ O \\ O \end{bmatrix}, B_{wx}^H = \begin{bmatrix} -1/L_F \cdot I & O \\ O & O \\ O & O \end{bmatrix} \\
C_x^H &= \left[I \; O \; O \right], D_x^H = \left[O \; -I \right]
\end{aligned}
\tag{7.12}
$$

where I is the identity matrix, O is the null matrix, ${G_x}^H$ is the matrix contains the harmonic components of G_x, which are written as G_{xn} for all positive and negative integer n.

Generally, the HSS model is in infinite dimension. However, the high order harmonic components are usually small and can be simplify neglected. Thus, a finite dimension model can be obtained with a truncation of matrix size.

7.3 H∞ Optimal Controller Design for TC*LC*-HAPF

From (7.9)–(7.12), the HSS model of the TC*LC*-HAPF can be built as a LTI system, which enables the optimal control design methods for the TC*LC*-HAPF. In this chapter, the HOC design method for the TC*LC*-HAPF is proposed.

The H_∞ controller design is to find an output feedback controller that stabilizes the TC*LC*-HAPF system and meets the H_∞ performance constraint:

$$
\|T_{wz}\|_\infty = \left\| D_x^H + C_x^H (sI - A_x^H) B_w^H \right\|_\infty < \gamma \tag{7.13}
$$

where T_{wz} is the closed-loop transfer function from W_x to Y_x, γ is a positive constant for the system H_∞ performance evaluation. To obtain the H_∞ controller satisfies (7.13), the following LMI problem [13] is formulated:

$$
\begin{bmatrix} A_{CL}^T P + OA_{CL} & PB_{CL} & C_{CL}^T \\ B_{CL}^T P & -\gamma I & D_{CL}^T \\ C_{CL} & D_{CL} & -\gamma I \end{bmatrix} < 0 \tag{7.14}
$$

$$P > 0 \tag{7.15}$$

where P is auxiliary variable, A_{CL}, B_{CL}, C_{CL}, and D_{CL} are the closed-loop HSS matrices which are given in the following:

$$A_{CL} = A_x^H + B_{ux}^H K_x^H C_x^H \quad B_{CL} = B_{wx}^H \tag{7.16}$$

$$C_{CL} = C_x^H + D_x^H K_x^H D_x^{HT} \quad D_{CL} = D_x^H \tag{7.17}$$

where ${K_x}^H$ is the control gain matrix. The LMI problem described in (7.14)–(7.17) can be efficiently solved by using the well-developed software packages [14].

From (7.13), the H_∞ performance index is represented by a constant γ, the smaller γ yields better system performance. To find a minimum γ constrained by the system dynamics, a γ-iteration algorithm is proposed, which is shown in Fig. 7.2. First of all, the data of HSS matrices (7.12) is collected, and γ is initialized as a small step value $\Delta\gamma$. Then the LMI problem (7.14)–(7.17) is solved with the initial value of γ. If the LMI problem is infeasible (that means there is no ${K_x}^H$ meets the conditions (7.14)–(7.17), the γ is updated by:

$$\gamma = \gamma + \Delta\gamma \tag{7.18}$$

This algorithm iteratively solves the LMI problem corresponding to the updated γ value, until there is a feasible solution. If the LMI problem has a feasible solution, then return the control gain matrix ${K_x}^H$ as the optimal control gain matrix ${K_{optx}}^H$. Therefore, the HOC is obtained as:

$$U_x = K_{optx}^H Y_x = \begin{bmatrix} \ddots & \vdots & \vdots & \vdots & ⋰ \\ \cdots & K_{optx-11} & K_{optx01} & K_{optx11} & \cdots \\ \cdots & K_{optx-12} & K_{optx02} & K_{optx12} & \cdots \\ \cdots & K_{optx-13} & K_{optx03} & K_{optx13} & \cdots \\ ⋰ & \vdots & \vdots & \vdots & \ddots \end{bmatrix} \begin{bmatrix} \vdots \\ Y_{x-1} \\ Y_{x0} \\ Y_{x1} \\ \vdots \end{bmatrix} \tag{7.19}$$

The HOC (7.19) in the HSS cannot be implemented directly, which should be transformed back into the time domain signals, which can be derived as:

$$u_x(t) = \sum_{n=-h}^{+h} U_{xn} e^{jn\omega_1 t} = \sum_{n=-h}^{+h} \sum_{m=-h}^{+h} K_{optxnm} Y_{xn} e^{jn\omega_1 t} \tag{7.20}$$

However, the negative harmonic components of Fourier series are proposed for mathematical completeness, which is not convenient for the practical implementation of (7.20). To address this concern, the following equivalent expression of (7.20) are used:

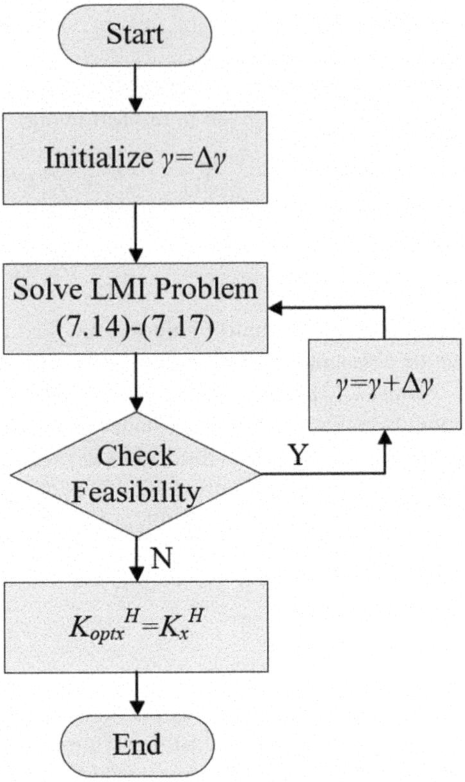

Fig. 7.2 Flow chart of the proposed γ iteration algorithm

$$y_{xn}(t) = Y_{xpn} cos(n\omega_1 t) + Y_{xqn} \sin(n\omega_1 t) = Y_{xn} e^{jn\omega_1 t} + Y_{x-n} e^{-jn\omega_1 t} \quad (7.21)$$

where $y_{xn}(t)$ is the n-order harmonic component of $y_x(t)$, Y_{xpn} and Y_{xqn} are active and reactive amplitudes components of $y_{xn}(t)$, which can be extracted by using band-pass filters (BPFs) [15]. Using the second order BPFs for each harmonic order with the zero-damping ratio, the s-domain transfer function $C_{H\infty}$(s) of the HOC can be derived as follows:

$$\begin{aligned} u_x(s) &= v_{invx}{}^{*}(s) = C_{H\infty}(s) y_x(s) \\ &= \sum_{m=0}^{h} K_{optx0m} + \sum_{n=1}^{h} \frac{2\sum_{m=0}^{h} K_{optxnm} s}{s + (n\omega_1)^2} \left[i_{cx}^{*}(s) - i_{cx}(s) \right] \end{aligned} \quad (7.22)$$

From (7.22), the resultant transfer function of the HOC is similar to the muti-resonant controller which has been widely adopted for the APF and HAPFs [16–19]. However, the conventional *s*-domain design of muti-resonant controller [17–19] cannot be directly applied to the nonlinear TC*LC*-HAPF, as the Laplace transform requires a linear system model. A linear approximation method of the TC*LC*-HAPF is recently proposed in [16], and thus MQPRC can be applied. However, the approximation model neglects the switching dynamic of thyristors and makes the MQPRC design not optimal. Compared to the existing muti-resonant controller design, the proposed HSS model without any approximation and the proposed optimization-based design guarantee the optimality of the controller parameters, and hence improve the control performance.

7.4 Overall Control Block Diagram

According to the previous works [7–12], the TC*LC* part and the active inverter part should be controlled coordinately. The overall control block diagram of the TC*LC*-HAPF is illustrated by Fig. 7.3, which consists of TC*LC* part control and active inverter part control. The TC*LC* part control is presented in Sect. 6.3.5 and will not be covered again in this chapter.

After performing all the calculation, the HOC will generate a voltage reference signal v_{invx}^{*} to compare with the fixed frequency triangular carrier wave for generating inverter trigger signals. The switching frequency of the switching pulse is 10 kHz which is same as the triangular carrier wave.

7.4.1 Simulation Results

In this section, simulation results of applying different controllers for the TC*LC*-HAPF are provided by using MATLAB/Simulink. The system, the TC*LC*-HAPF and the load parameters are provided in Table 7.1, in which the DC-link voltage is modelled by a DC voltage source to simplify the verification of the validity and effectiveness of the HOC for the TC*LC*-HAPF. The simulation results are presented in Fig. 7.4 and summarized in Table 7.2.

Figure 7.4a, c and Table 7.2 show the TC*LC*-HAPF compensating performance with the MQPRC under inductive and capacitive loads compensation. The steady-state compensation performance is satisfactory for both load conditions. In steady-state, the worst-case THDi_{sx} can be reduced to 2.03% and 2.13% for inductive and capacitive loads, respectively. However, it takes a long transient time from the TC*LC*-HAPF starting operation to the steady-state (65 and 57 ms for inductive and capacitive loads), which is because the gains design is not optimized and the proportional gain is also the same for the fundamental and all considered harmonics order, but our

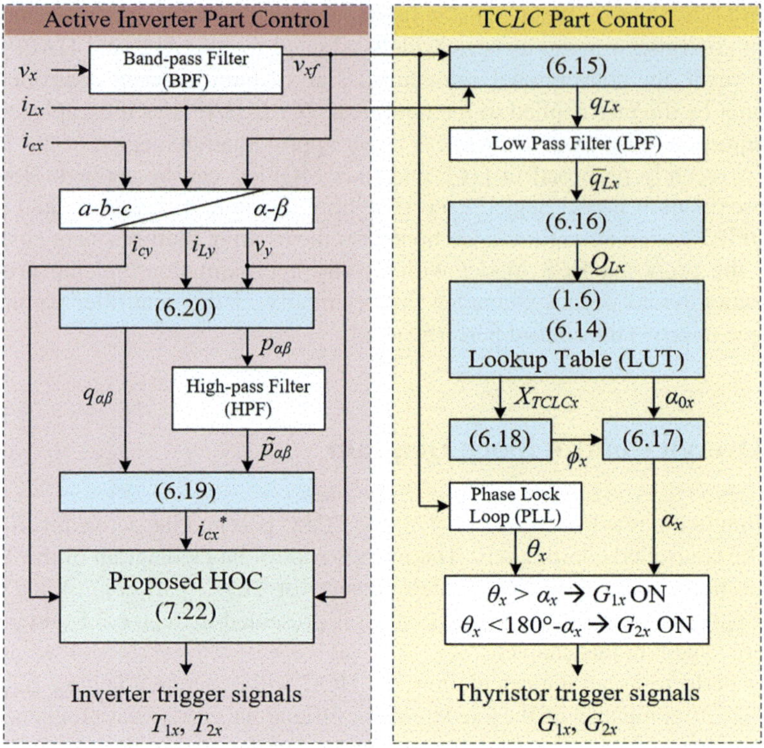

Fig. 7.3 Overall control block diagram of the TCLC-HAPF with the proposed HOC

Table 7.1 Parameters of system, TCLC-HAPF, loads for simulation and experiment

	Parameters	Value
Grid parameters	v_{gx}, f_g	110 V, 50 Hz
Inverter parameters	V_{dc}, f_{sw}	60 V, 10 kHz
TCLC parameters	L_c, L_{PF}, C_{PF}	5 mH, 30 mH, 160 μF
Sampling time	T_s	100 μs
Non-linear load	R_{NL}, L_{NL}	50 Ω, 30 mH
Inductive load	R_{LL}, L_{LL}	15 Ω, 70 mH
Capacitive load	R_{LL}, C_{LL}	25 Ω, 140 μF

proposed HOC design provides the optimized gains for the fundamental and each of the harmonics order.

In Fig. 7.4b, d, and Table 7.2, the TCLC-HAPF compensating performance with the proposed HOC is demonstrated. From Fig. 7.4b and d, the steady-state compensation performance is better than the MQPRC under both load conditions, which reduces the worst-case THDi_{sx} to 1.83 and 1.91% for inductive and capacitive loads.

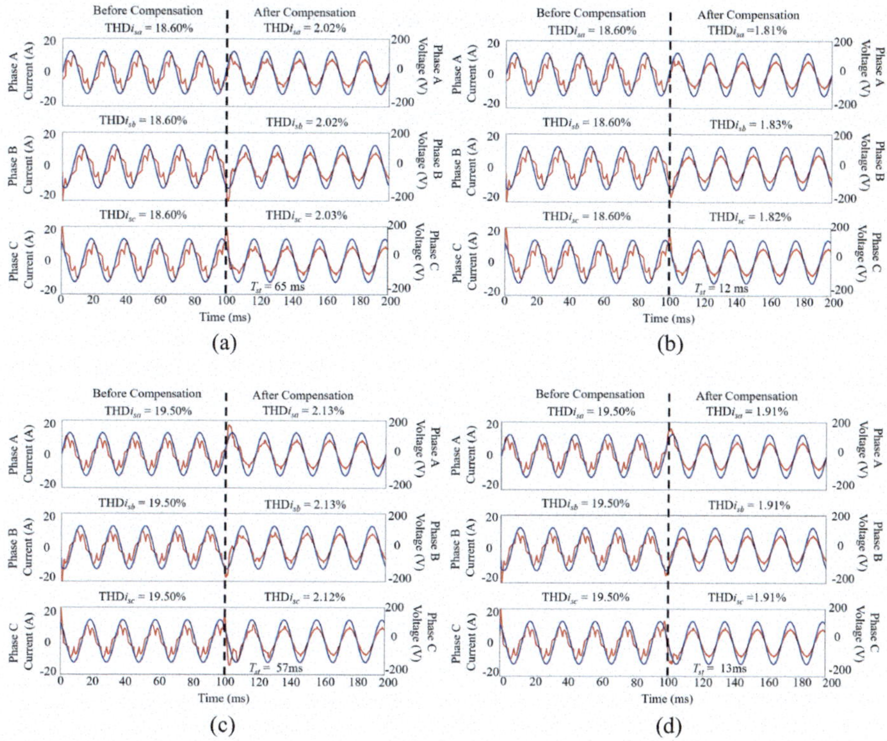

Fig. 7.4 Simulation results of TC*LC*-HAPF compensating performance by using different controllers under inductive and capacitive load conditions: **a** MQPRC under inductive load; **b** proposed HOC under inductive load; **c** MQPRC under capacitive load; **d** proposed HOC under capacitive load

Table 7.2 Simulation results before and after TC*LC*-HAPF compensation

		P_{sx} (W)	Q_{sx} (var)	PF	T_{st} (ms)	Worst THDi_{sx} (%)
Inductive Loads	Before Compensation	650	450	0.78	/	18.6
	MQPRC	660	5	0.99	65	2.03
	HOC	660	3	0.99	12	1.83
Capacitive Loads	Before Compensation	750	−250	0.93	/	19.5
	MQPRC	760	−4	0.99	57	2.13
	HOC	760	−2	0.99	13	1.91

Moreover, the transient response is much faster than the MQPRC, which are 12 ms and 13 ms for inductive load and capacitive load, respectively.

In conclusion, with the help of HSS modeling and optimization-based control design, the proposed HOC provides superior steady-state and transient performance simultaneously compared with the state-of-the-art MQPRC [16].

7.5 Experimental Results

To verify the simulation results, a three-phase three-wire 110V–5kVA TC*LC*-HAPF experimental prototype was built in the laboratory as discussed in Chap. 6. Table 7.1 shows the experimental setting of the system and controllers' parameters. The experimental results are presented in Fig. 7.5 and summarized in Table 7.3.

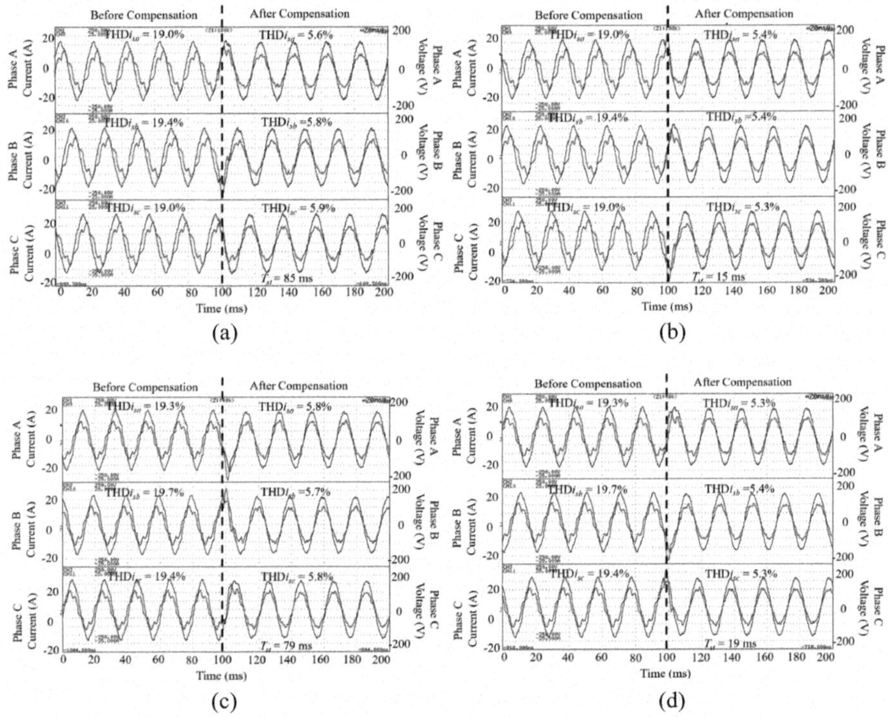

Fig. 7.5 Experimental results of TC*LC*-HAPF compensating performance by using different controller under inductive and capacitive load conditions: **a** MQPRC under inductive load; **b** proposed HOC under inductive load; **c** MQPRC under capacitive load; **d** proposed HOC under capacitive load

Table 7.3 Experimental results before and after TC*LC*-HAPF compensation

		P_{sx} (W)	Q_{sx} (var)	PF	T_{st} (ms)	Worst THDi_{sx} (%)
Inductive Loads	Before Compensation	610	450	0.78	/	19.4
	MQPRC	660	10	0.99	85	5.9
	HOC	660	10	0.99	15	5.4
Capacitive Loads	Before Compensation	740	−250	0.93	/	19.7
	MQPRC	770	−10	0.99	79	5.8
	HOC	770	−10	0.99	19	5.4

Figure 7.5a, c and Table 7.3 show the TC*LC*-HAPF compensating performance with the MQPRC under inductive and capacitive loads compensation. The steady-state compensation performance is satisfactory for both load conditions. In steady-state, the worst-case THDi_{sx} can be reduced to 5.9% and 5.8% for inductive and capacitive loads, respectively. However, it takes a long transient time from the TC*LC*-HAPF starting operation to the steady-state (85 and 79 ms for inductive and capacitive loads), which is because the gains design is not optimized, and the proportional gain is also the same for the fundamental and all considered harmonics order.

In Fig. 7.5b, d and Table 7.3, the TC*LC*-HAPF compensating performance with the proposed HOC is demonstrated. From Fig. 7.5b, d, the steady-state compensation performance is better than the MQPRC for inductive and capacitive loads, which reduces the worst-case THDi_{sx} to 5.4% under both load conditions. Moreover, the transient response is much faster than the MQPRC, which are 15 ms and 19 ms for inductive load and capacitive load, respectively.

Therefore, both simulation and experimental results verify that the proposed HOC based on HSS and LMI provides superior steady-state and transient performance simultaneously compared to the state-of-the-art MQPRC proposed in [16].

In conclusion, as the gains design of the MQPRC is not optimized and the proportional gain is designed to be the same for both fundamental and all considered harmonics order, the transient response of the MQPRC is slow. On the contrary, our proposed HOC design provides the optimized gains for both fundamental and each of the harmonics order, which results in superior steady-state and transient performances.

7.6 Summary

To address the nonlinearity and improve the performance of the TC*LC*-HAPF, an HOC design based on HSS modeling for the TC*LC*-HAPF is proposed in this chapter. First, the HSS model in LTI form of the TC*LC*-HAPF is proposed to address the system nonlinearity issue. Then, the HOC design based on the LMI is proposed, which can improve the system performance. Finally, simulation and experimental results are provided to verify the effectiveness and performance of the HOC for the TC*LC*-HAPF in comparison with the recent developed state-of-the-art MQPRC, which shows superior steady-state and transient control performances.

In addition, with the help of the proposed HSS model, the linear control designs are unlocked for the nonlinear TC*LC*-HAPF. Therefore, more recent works for linear optimal control proposed by the control system society could be developed on the TC*LC*-HAPF, which can potentially further improve the system performance and reduce the algorithm complexity.

References

1. S. Rahmani, A. Hamadi, K. Al-Haddad, L.A. Dessaint, A combination of shunt hybrid power filter and thyristor-controlled reactor for power quality. IEEE Trans. Ind. Electron. **61**(5), 2152–2164 (2014)
2. S. Skogestad, I. Postlethwaite, *Multivariable Feedback Control: Analysis and Design*, vol. 2 (Wiley, New York, 2007)
3. J.R.C. Orillaza, A.R. Wood, Harmonic state-space model of a controlled TCR. IEEE Trans. Power Deliv. **28**(1), 197–205 (2013)
4. G.N. Love, A.R. Wood, Harmonic state space model of power electronics, in *2008 13th International Conference on Harmonics and Quality of Power* (2008), pp. 1–6
5. J. Kwon, X. Wang, F. Blaabjerg, C.L. Bak, V.-S. Sularea, C. Busca, Harmonic interaction analysis in a grid-connected converter using harmonic state-space (HSS) modeling. IEEE Trans. Power Electron. **32**(9), 6823–6835 (2017)
6. N.M. Wereley, S.R. Hall, Linear time periodic systems: transfer function, poles, transmission zeroes and directional properties, in *1991 American Control Conference* (1991), pp. 1179–1184
7. L. Wang, C.-S. Lam, M.-C. Wong, A hybrid-STATCOM with wide compensation range and low DC-link voltage. IEEE Trans. Ind. Electron. **63**(6), 3333–3343 (2016)
8. C.-S. Lam, L. Wang, S.-I. Ho, M.-C. Wong, Adaptive thyristor-controlled LC-hybrid active power filter for reactive power and current harmonics compensation with switching loss reduction. IEEE Trans. Power Electron. **32**(10), 7577–7590 (2016)
9. L. Wang, C.-S. Lam, M.-C. Wong, Unbalanced control strategy for a thyristor-controlled LC-coupling hybrid active power filter in three-phase three-wire systems. IEEE Trans. Power Electron. **32**(2), 1056–1069 (2016)
10. L. Wang, C.-S. Lam, M.-C. Wong, Selective compensation of distortion, unbalanced and reactive power of a thyristor-controlled LC-coupling hybrid active power filter (TCLC-HAPF). IEEE Trans. Power Electron. **32**(12), 9065–9077 (2017)
11. L. Wang, C.-S. Lam, M.-C. Wong, Analysis, control, and design of a hybrid grid-connected inverter for renewable energy generation with power quality conditioning. IEEE Trans. Power Electron. **33**(8), 6755–6768 (2017)
12. L. Wang, C.-S. Lam, M.-C. Wong, Minimizing inverter capacity design and comparative performance evaluation of SVC-coupling hybrid active power filters. IEEE Trans. Power Electron. **34**(2), 1227–1242 (2018)
13. S. Boyd, L. El Ghaoui, E. Feron, V. Balakrishnan, *Linear Matrix Inequalities in System and Control Theory*. SIAM (1994)
14. D.-W. Gu, P.H. Petkov, M.M. Konstantinov, *Robust Control Design With MATLAB* (Springer, New York, NY, USA, 2005)

15. J.C. Willems, S. Hara, Y. Ohta, H. Fujioka (eds.), *Perspectives in Mathematical System Theory, Control, and Signal Processing*, vol. 398 (Springer Berlin Heidelberg, Berlin, Heidelberg, 2010)
16. W.-K. Sou, C.-W. Chao, C. Gong, C.-S. Lam, C.-K. Wong, Analysis, design and implementation of multi-quasi-proportional-resonant controller for thyristor-controlled LC-coupling hybrid active power filter (TCLC-HAPF). IEEE Trans. Ind. Electron., in Early Access
17. A.G. Yepes, F.D. Freijedo, Ó. Lopez, J. Doval-Gandoy, High-performance digital resonant controllers implemented with two integrators. IEEE Trans. Power Electron. **26**(2), 563–576 (2011)
18. Z. He, F. Ma, M. Li, Q. Xu, F. Zhou, Y. Yue, Multi-functional hybrid active power converter and its industrial application for electrolytic copper-foil. IET Gener. Transm. Distrib. **10**(16), 3985–3994 (2016)
19. C. Xie, X. Zhao, K. Li, J. Zou, J.M. Guerrero, A new tuning method of multiresonant current controllers for grid-connected voltage source converters. IEEE Trans. Emerg. Sel. Top. Power Electron. **7**(1), 458–466 (2019)

Chapter 8
Finite-Set Model Predictive Control for TC*LC*-HAPF

8.1 Introduction

In this chapter, a finite-set model predictive control (FS-MPC) with coupling capacitor voltage observer scheme is proposed for TC*LC*-HAPF. It ensures the fast transient response, low steady-state error and good robustness under both inductive and capacitive load situations. Generally, the implementation of FS-MPC for TC*LC*-HAPF requires sensing of coupling capacitor voltage and thyristor-controlled reactor current in addition to sensing load voltage, load current and compensation current, which increases the system complexity and cost. To relax this problem, the simplified predictive model is first proposed, thus reduce the required sensing signals and system order. Then, a coupling capacitor voltage observer is utilized to reduce the hardware costs, which can obtain a comparable performance with directly sensed capacitor voltage. The delay compensation is also considered for the control delay in practical realization. Finally, the effectiveness and performance of the FS-MPC with coupling capacitor voltage observer for TC*LC*-HAPF are verified by simulated and experimental results.

The original contribution of this chapter is summarized as follows:

(1) Deduce the discrete-time predictive model for the TC*LC*-HAPF, so that FS-MPC can be designed and implemented accordingly;
(2) Simplify the system model of TC*LC*-HAPF for the ease of implementing FS-MPC;
(3) Propose a FS-MPC with coupling capacitor voltage observer for the TC*LC*-HAPF, which provides a fast transient response, low steady-state error and good robustness under inductive and capacitive load conditions;
(4) Provide the parameter design of coupling capacitor voltage observer to ensure the stability and transient performance;
(5) Verify the effectiveness and performance of the proposal by simulation and experimental results.

C.-S. Lam et al., *Advanced Controllers Design for Hybrid Active Power Filters*,
https://doi.org/10.1007/978-981-97-6847-9_8

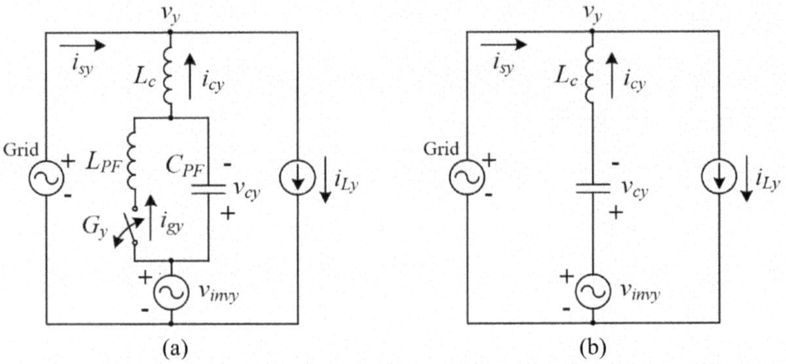

Fig. 8.1 Equivalent circuit of TCLC-HAPF in α-β frame. **a** Real model. **b** Simplified model

8.2 Modeling of TCLC-HAPF

From the circuit configuration of the three-phase three-wire TCLC-HAPF shown in Fig. 1.5, the equivalent circuit in complex α-β frame of the TCLC-HAPF is shown in Fig. 8.1a, where subscript 'y' denotes frame α and β. Based on this, the system dynamic model can be described by the following equations, where v_y, v_{cy} are the load and coupling capacitor voltage, i_{gy} is the thyristor-controlled reactor current and G_y is on–off state of a pair of anti-parallel thyristors.

$$\begin{cases} L_c \frac{di_{cy}}{dt} = v_{invy} - v_y - v_{cy} \\ C_{PF} \frac{dv_{cy}}{dt} = i_{cy} - i_{gy} \\ L_{PF} \frac{di_{ty}}{dt} = G_y v_{cy} \end{cases} \tag{8.1}$$

8.3 Design of Finite-Set Model Predictive Control for TCLC-HAPF

8.3.1 Predictive Model Construction

(A) ***Real Model***: According to (8.1), the continuous-time state-space equation of the TCLC-HAPF is written as below.

$$\underbrace{\frac{d}{dt}\begin{bmatrix} i_{cy} \\ v_{cy} \\ i_{gy} \end{bmatrix}}_{\dot{\mathbf{x}}'(t)} = \underbrace{\begin{bmatrix} 0 & -\frac{1}{L_c} & 0 \\ \frac{1}{C_{PF}} & 0 & -\frac{1}{C_{PF}} \\ 0 & \frac{T_y}{L_{PF}} & 0 \end{bmatrix}}_{\mathbf{A_t}} \underbrace{\begin{bmatrix} i_{cy} \\ v_{cy} \\ i_{gy} \end{bmatrix}}_{\mathbf{x}'(t)} + \underbrace{\begin{bmatrix} \frac{1}{L_c} \\ 0 \\ 0 \end{bmatrix}}_{\mathbf{B_t}} \underbrace{\begin{bmatrix} v_{invy} - v_y \end{bmatrix}}_{\mathbf{u}(t)} \tag{8.2}$$

By applying zero-order hold approach, the generalized form of typical discrete-time predictive model of the TCLC-HAPF is shown as (8.3).

$$\underbrace{\begin{bmatrix} i_{cy}(k+1) \\ v_{cy}(k+1) \\ i_{gy}(k+1) \end{bmatrix}}_{\mathbf{x}'(k+1)} = \mathbf{\Phi}_{\mathbf{t}} \underbrace{\begin{bmatrix} i_{cy}(k) \\ v_{cy}(k) \\ i_{gy}(k) \end{bmatrix}}_{\mathbf{x}'(k)} + \mathbf{\Gamma}_{\mathbf{t}} \underbrace{\left[v_{invy}(k) - v_y(k)\right]}_{\mathrm{u}(k)} \tag{8.3}$$

where $\Phi_{\mathrm{t}} = e^{\mathrm{A_t}T_s}$, $\Gamma_{\mathrm{t}} = \int_0^{T_s} e^{\mathrm{A_t}\tau}\mathrm{B_t}d\tau$, and T_s is the sampling period. The matrices $\mathbf{\Phi}_{\mathbf{t}}$ and $\mathbf{\Gamma}_{\mathbf{t}}$ is a 3×3 matrix and 3×1 matrix which are shown in below, where $\omega_n = \frac{1}{\sqrt{\frac{L_c C_{PF} L_{PF}}{L_{PF}+L_c T_y}}}$.

$$\begin{cases} \Phi_{\mathrm{t}} = \begin{bmatrix} \frac{L_{PF}\cos(\omega_n T_s)+L_c G_y}{L_{PF}+L_c G_y} & -\frac{1}{\omega_n L_c}\sin(\omega_n T_s) & \frac{1}{L_c C_{PF}\omega_n^2}[1-\cos(\omega_n T_s)] \\ \frac{1}{\omega_n C_{PF}}\sin(\omega_n T_s) & \cos(\omega_n T_s) & -\frac{1}{\omega_n C_{PF}}\sin(\omega_n T_s) \\ \frac{G_y}{L_{PF}C_{PF}\omega_n^2}[1-\cos(\omega_n T_s)] & \frac{G_y}{L_{PF}\omega_n}\sin(\omega_n T_s) & \frac{L_c G_y\cos(\omega_n T_s)+L_{PF}}{L_{PF}+L_c G_y} \end{bmatrix} \\ \Gamma_{\mathrm{t}} = \begin{bmatrix} \frac{1}{L_{PF}+L_c T_y}\left[\frac{L_{PF}}{\omega_n L_c}\sin(\omega_n T_s) + G_y T_s\right] \\ 1-\cos(\omega_n T_s) \\ \frac{G_y}{L_{PF}+L_c G_y}\left[T_s - \frac{1}{\omega_n}\sin(\omega_n T_s)\right] \end{bmatrix} \end{cases} \tag{8.4}$$

Obviously, (8.1)–(8.3) are representing a 3rd order system. To implement the FS-MPC for the TCLC-HAPF by using the real model (8.3), the signals of the compensating current $i_{cy}(k)$, coupling capacitor voltage $v_{cy}(k)$, load voltage $v_y(k)$, thyristor-controlled reactor current $i_{gy}(k)$ and on–off state of thyristors $G_y(k)$ are required ($G_y(k)$ can be obtained from TCLC part control). Theoretically, all the signals should be directly sensed and transformed into the complex α-β frame. However, this approach will greatly increase the hardware and maintenance costs, and the system reliability will also decrease.

From the previous study, at least three signals need to be sensed for controlling the TCLC-HAPF, which are v_x, i_{cx} and i_{Lx} [1–4]. If the FS-MPC is designed without using any additional sensing circuit, the rest of the signals (v_{cx}, i_{gx}) can only be estimated by using observers. Nevertheless, applying the typical FS-MPC model for the TCLC-HAPF is still challenging, because the coupling TCLC part is a nonlinear switching system that makes the design of the observer complicated. In order to implement the FS-MPC for the TCLC-HAPF more easily, a simplified predictive control model for the TCLC-HAPF is proposed hereafter.

(B) ***Proposed Simplified Model***: From Fig. 8.1a, since the capacitor C_{PF} is in parallel with the thyristor-controlled reactor L_{PF}, the coupling capacitor voltage v_{cy} is equal to the voltage of the thyristor-controlled reactor. Hence, the TCLC part circuit can be simplified as shown in Fig. 8.1b. From Fig. 8.1b, the simplified TCLC-HAPF system dynamic model is derived as (8.5) and the corresponding continuous-time state-space equation is given in (8.6).

$$\begin{cases} L_c\frac{di_{cy}}{dt} = v_{invy} - v_y - v_{cy} \\ C_{PF}\frac{dv_{cy}}{dt} = i_{cy} \end{cases} \tag{8.5}$$

$$\underbrace{\frac{d}{dt}\begin{bmatrix} i_{cy} \\ v_{cy} \end{bmatrix}}_{\dot{\mathrm{x}}(t)} = \underbrace{\begin{bmatrix} 0 & -\frac{1}{L_c} \\ \frac{1}{C_{PF}} & 0 \end{bmatrix}}_{\mathrm{A}} \underbrace{\begin{bmatrix} i_{cy} \\ v_{cy} \end{bmatrix}}_{\mathrm{x}(t)} + \underbrace{\begin{bmatrix} \frac{1}{L_c} \\ 0 \end{bmatrix}}_{\mathrm{B}} \underbrace{\begin{bmatrix} v_{invy} - v_y \end{bmatrix}}_{\mathrm{u}(t)} \tag{8.6}$$

According to (8.6), the discrete-time predictive control model of the TCLC-HAPF is rewritten as (8.7), in which the coefficient matrices $\mathbf{\Phi}$ and $\mathbf{\Gamma}$ are formulated as (8.8), and $\omega_n = \frac{1}{\sqrt{L_c C_{PF}}}$.

$$\underbrace{\begin{bmatrix} i_{cy}(k+1) \\ v_{cy}(k+1) \end{bmatrix}}_{\mathrm{x}(k+1)} = \underbrace{\begin{bmatrix} \Phi_{11} & \Phi_{12} \\ \Phi_{21} & \Phi_{22} \end{bmatrix}}_{\Phi} \underbrace{\begin{bmatrix} i_{cy}(k) \\ v_{cy}(k) \end{bmatrix}}_{\mathrm{x}(k)} + \underbrace{\begin{bmatrix} \Gamma_1 \\ \Gamma_2 \end{bmatrix}}_{\Gamma} \underbrace{\begin{bmatrix} v_{invy}(k) - v_y(k) \end{bmatrix}}_{\mathrm{u}(k)} \tag{8.7}$$

$$\begin{cases} \Phi = \begin{bmatrix} \cos(\omega_n T_s) & -\frac{1}{\omega_n L_c}\sin(\omega_n T_s) \\ \frac{1}{\omega_n C_{PF}}\sin(\omega_n T_s) & \cos(\omega_n T_s) \end{bmatrix} \\ \Gamma = \begin{bmatrix} \frac{1}{\omega_n L_c}\sin(\omega_n T_s) \\ 1 - \cos(\omega_n T_s) \end{bmatrix} \end{cases} \tag{8.8}$$

It can be seen clearly that (8.5)–(8.7) are representing a 2nd order system. Compared with (8.1)–(8.3), the order of the predictive control model is reduced by this simplification, thereby making the FS-MPC easier to be implemented for the TCLC-HAPF. From (8.7), the signal $v_{cy}(k)$ is still required for predicting the $i_{cy}(k+1)$ and $v_{cy}(k+1)$. In this work, $v_{cy}(k)$ is estimated by using the coupling capacitor voltage observer to lower the hardware sensing cost.

8.3.2 *Coupling Capacitor Voltage Estimation*

(A) ***Coupling Capacitor Voltage Observer***: A full-order coupling capacitor voltage v_{cy} observer is adopted for estimating the v_{cy} after considering the stability, simpleness and noise immunity of the observer in practical applications [5]. It is expressed as:

$$\frac{d\hat{i}_{cy}}{dt} = \frac{1}{L_c}(v_{invy} - v_y - \hat{v}_{cy}) + k_1 e_y \tag{8.9}$$

$$\frac{d\hat{v}_{cy}}{dt} = \frac{1}{C_{PF}}\hat{i}_{cy} + k_2 e_y \tag{8.10}$$

where the symbol '^' represents the estimated value, k_1 and k_2 are the observer gains and $e_y = i_{cy} - \hat{i}_{cy}$ is the estimation error of compensating current.

Therefore, the v_{cy} can be estimated by using the load voltage v_y and compensating current i_{cy}. Moreover, the discrete form of the observer in (8.9) and (8.10) can be derived by using the forward Euler method, which can be expressed as:

$$\hat{i}_{cy}(k+1) = \hat{i}_{cy}(k) + T_s\left(\frac{1}{L_c}\left(v_{invy}(k) - v_y(k) - \hat{v}_{cy}(k)\right) + k_1 e_y(k)\right) \tag{8.11}$$

$$\hat{v}_{cy}(k+1) = \hat{v}_{cy}(k) + T_s\left(\frac{1}{C_{PF}}\hat{i}_{cy}(k) + k_2 e_y(k)\right) \tag{8.12}$$

where $v_{invy}(k)$ is the optimal inverter voltage.

(B) ***Observer Gains Design***: The observer gains k_1 and k_2 in (8.11) and (8.12) can be designed by using the pole placement method [6]. The design procedure is given in the following.

Based on the discrete equation shown in (8.11) and (8.12), the discrete-time state-space model of the observer is written as (8.13), where **L**, **M**, and **N** are the observer matrices.

$$\begin{aligned}\underbrace{\begin{bmatrix}\hat{i}_{cy}(k+1)\\ \hat{v}_{cy}(k+1)\end{bmatrix}}_{\hat{\mathrm{x}}(k+1)} &= \underbrace{\begin{bmatrix}1 & -\frac{T_s}{L_c}\\ \frac{T_s}{C_{PF}} & 1\end{bmatrix}}_{\mathrm{M}}\underbrace{\begin{bmatrix}\hat{i}_{cy}(k)\\ \hat{v}_{cy}(k)\end{bmatrix}}_{\hat{\mathrm{x}}(k)} + \underbrace{\begin{bmatrix}k_1T_s & 0\\ k_2T_s & 0\end{bmatrix}}_{\mathrm{L}}\underbrace{\begin{bmatrix}i_{cy}(k) - \hat{i}_{cy}(k)\\ v_{cy}(k) - \hat{v}_{cy}(k)\end{bmatrix}}_{\mathrm{x}(k)-\hat{\mathrm{x}}(k)}\\ &+ \underbrace{\begin{bmatrix}\frac{T_s}{L_c}\\ 0\end{bmatrix}}_{\mathrm{N}}\underbrace{\left[v_{invy}(k) - v_y(k)\right]}_{\mathrm{u}(k)}\\ &= (\mathrm{M}-\mathrm{L})\hat{\mathrm{x}}(k) + \mathrm{L}\mathrm{x}(k) + \mathrm{N}\mathrm{u}(k)\end{aligned} \tag{8.13}$$

The design of the observer depends on two requirements: stability and dynamic response. To ensure the stability, the eigenvalues (i.e., poles) of (8.13) should be placed within the unit circle of the z-plane. In this work, both eigenvalues of (8.13) are set to 0 (the origin of the z-plane) to obtain a deadbeat response [6].

Hence, the observer can achieve a stable and fast dynamic response simultaneously. The corresponding eigenpolynomial and poles setting condition can be expressed as (8.14), where **I** is a 2×2 identity matrix.

$$\begin{aligned}\det(z\mathrm{I} - \mathrm{M} + \mathrm{L}) &= z^2 + (k_1T_s - 2)z + 1 - k_1T_s + \frac{T_s^2}{L_cC_{PF}} - \frac{k_2T_s^2}{L_c}\\ &= (z-0)^2\end{aligned} \tag{8.14}$$

After comparing the coefficients of (8.14), the observer gains k_1 and k_2 can be calculated as (8.15).

$$\begin{cases} k_1 = \frac{2}{T_s} \\ k_2 = \frac{1}{C_{PF}} - \frac{L_c}{T_s} \end{cases} \tag{8.15}$$

8.3.3 Delay Compensation

In practical application, there always exists a one-sampling-time delay due to the nature of digital control which affects the system performance [7]. Therefore, the delay compensation should be considered when designing a FS-MPC for the TCLC-HAPF. In this work, a two-step prediction method [8] is used to address the time delay problem. The objective of this method is to increase the prediction horizon from $(k + 1)$th to $(k + 2)$th. The following part will present the implementation of the delay compensation.

First, the $(k + 1)$th compensating current $i_{cy}(k + 1)$ and coupling capacitor voltage $v_{cy}(k + 1)$ are predicted by using the simplified predictive control model (8.7). Then, $i_{cy}(k + 2)$ is predicted by using (8.7) again with $i_{cy}(k + 1)$ and $v_{cy}(k + 1)$ predicted in previous step, which can be expressed as (8.16), where $v_{invy}(k + 1)$ is the optimal inverter voltage to be evaluated. In addition, $v_y(k + 1)$ can be replaced by $v_y(k)$ because the system grid period is much greater than the sampling time [7].

$$i_{cy}(k+2) = \Phi_{11} i_{cy}(k+1) + \Phi_{12} v_{cy}(k+1) + \Gamma_1 \big(v_{invy}(k+1) - v_y(k+1)\big) \tag{8.16}$$

However, $v_{cy}(k + 1)$ can be obtained by using (8.12) without using the predictive control model because both $\hat{v}_{cy}(k)$ and $\hat{v}_{cy}(k + 1)$ can be estimated. For the compensating current $i_{cy}(k + 1)$, it can also be obtained either by using the simplified predictive control model (8.7) or the observer (8.11). As the $i_{cy}(k)$ can be obtained by directly sensing of $i_{cx}(k)$ with the help of coordinate transformation, we chose the simplified model (8.7) for the delay compensation in this work. According to the above information, $i_{cy}(k + 1)$ is now obtained as (8.17). Correspondingly, $i_{cy}(k + 2)$ can be obtained as (8.18) with the help of (8.12), (8.16) and (8.17).

$$i_{cy}(k+1) = \Phi_{11} i_{cy}(k) + \Phi_{12} \hat{v}_{cy}(k) + \Gamma_1 \big(v_{invy}(k) - v_y(k)\big) \tag{8.17}$$

$$i_{cy}(k+2) = \Phi_{11} i_{cy}(k+1) + \Phi_{12} \hat{v}_{cy}(k+1) + \Gamma_1 \big(v_{invy}(k+1) - v_y(k)\big) \tag{8.18}$$

8.3.4 Cost Function and Optimal Vector Selection

To facilitate the design, the cost function chosen is shown in (8.19). Its objective is to minimize the compensating current tracking error, thus minimizing the THD*i* and reactive power.

$$J = (i_{c\alpha}{}^{*}(k+2) - i_{c\alpha}(k+2))^2 + (i_{c\beta}{}^{*}(k+2) - i_{c\beta}(k+2))^2 \tag{8.19}$$

where $i_{c\alpha}$ and $i_{c\beta}$ are the compensating current in α and β frame, respectively, and the notation '*' denotes the reference value. The reference $i_{cy}{}^{*}(k+2)$ is calculated by using the second order Lagrange extrapolation method [9], which is shown in (8.20).

$$i_{cy}{}^{*}(k+2) = 3i_{cy}{}^{*}(k) - 2i_{cy}{}^{*}(k-1) \tag{8.20}$$

Finally, substituting (8.18) into (8.19) for rolling optimization to obtain the inverter trigger signals. The working principle of the rolling optimization is illustrated as in Fig. 8.2. By enumerating all possible inverter voltage vector $v_{invy}(k+1)$ in (8.18), the optimal one that can generate the optimal $i_{cy}(k+2)$ to minimize (8.19) will be selected and then applied for the active inverter part. The relationship between the inverter voltage vector v_{invy} and the corresponding switching states S_x of each phase is listed in Table 8.1 [10]. Moreover, S_x are given as (8.21), where T_{1x} and T_{2x} are the inverter trigger signals of each phase.

$$S_x = \begin{cases} 1, & T_{1x}\ \text{ON},\ T_{2x}\ \text{OFF} \\ 0, & T_{1x}\ \text{OFF},\ T_{2x}\ \text{ON} \end{cases} \tag{8.21}$$

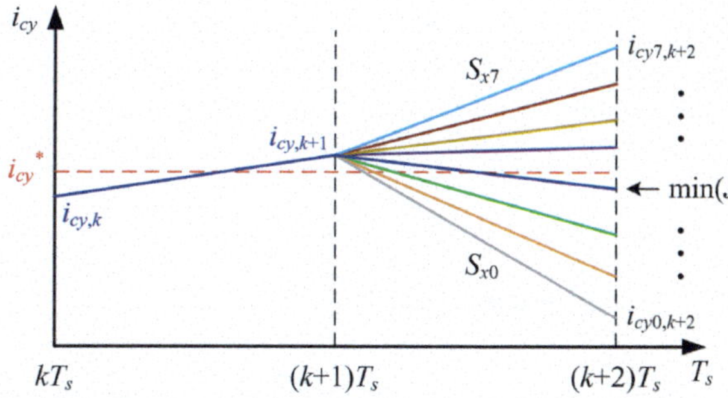

Fig. 8.2 Working principle of rolling optimization for FS-MPC

Table 8.1 Relationship between voltage vectors and switching states

Index	Switching State S_x			Voltage Vector v_{invy}	
	S_a	S_b	S_c	$v_{inv\alpha}$	$v_{inv\beta}$
0	0	0	0	0	0
1	1	0	0	$2V_{dc}/3$	0
2	1	1	0	$V_{dc}/3$	$\sqrt{3}V_{dc}/3$
3	0	1	0	$-V_{dc}/3$	$\sqrt{3}V_{dc}/3$
4	0	1	1	$-2V_{dc}/3$	0
5	0	0	1	$-V_{dc}/3$	$-\sqrt{3}V_{dc}/3$
6	1	0	1	$V_{dc}/3$	$-\sqrt{3}V_{dc}/3$
7	1	1	1	0	0

8.4 Overall Control Block Diagram

The comprehensive control block diagram of the TC*LC*-HAPF is illustrated by Fig. 8.3, which consists of TC*LC* part control and active inverter part control. The TC*LC* part control is presented in Sect. 6.3.5 and will not be covered again in this chapter.

The active inverter part is to compensate for the harmonic currents and the remaining load reactive power after the TC*LC* part compensation. Hence, the reference compensating current is calculated through the instantaneous *p*-*q* theory [11] as

$$\begin{bmatrix} i_{c\alpha}{}^* \\ i_{c\beta}{}^* \end{bmatrix} = \frac{1}{v_\alpha^2 + v_\beta^2} \begin{bmatrix} v_\alpha & -v_\beta \\ v_\beta & v_\alpha \end{bmatrix} \begin{bmatrix} \tilde{p}_{\alpha\beta} \\ q_{\alpha\beta} \end{bmatrix} \tag{8.22}$$

where $\tilde{p}_{\alpha\beta}$ and $q_{\alpha\beta}$ are three-phase instantaneous load harmonic power and reactive power. The former one is obtained with the help of the active power $p_{\alpha\beta}$ and the high-pass filter (HPF). The $p_{\alpha\beta}$ and $q_{\alpha\beta}$ is obtained by (6.20).

After getting the reference current, the proposed FS-MPC with coupling capacitor voltage observer will generate the inverter trigger signals T_{1x} and T_{2x} of each phase.

8.5 Simulation Results

The simulation studies for verifying the performance of the proposed FS-MPC with coupling capacitor voltage observer for the three-phase three-wire TC*LC*-HAPF are carried out by using MATLAB/Simulink. Table 8.2 lists the system parameters for simulation. To ease the verification of the proposal, a DC voltage source is used for providing the DC-link voltage. The simulation results are presented in Figs. 8.4, 8.5, 8.6, 8.7, 8.8, and 8.9 and Table 8.3. Note that the settling time T_s are obtained

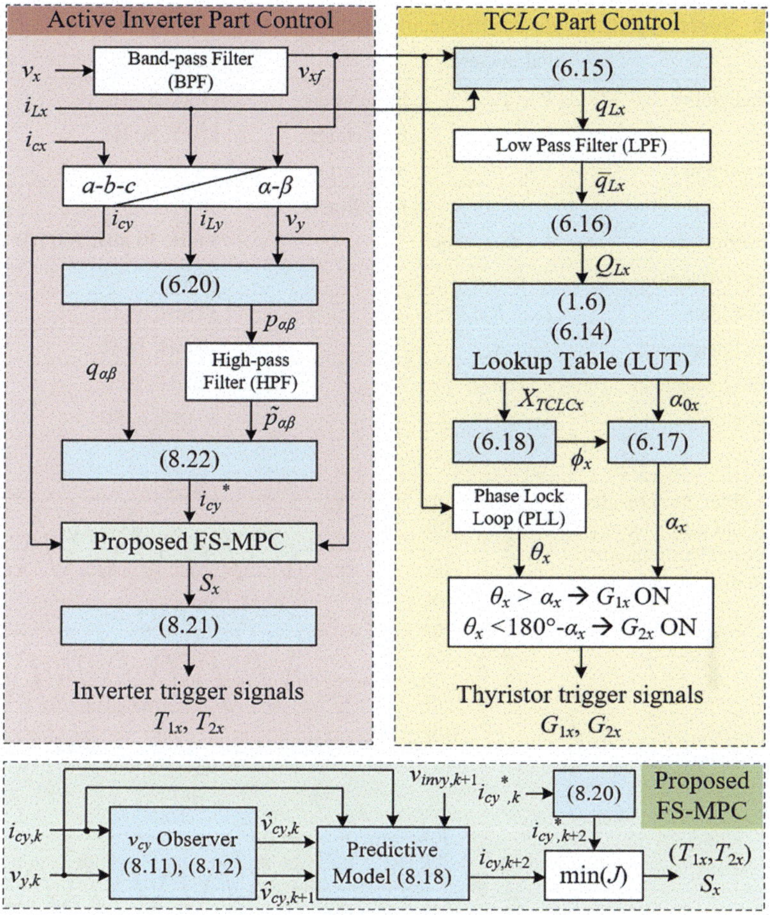

Fig. 8.3 Overall control block diagram of TCLC-HAPF for FS-MPC

when switching on the TC*LC*-HAPF with loads already connected, and the THD*i* are obtained when the system reaches the steady-state.

First, the simulation results of the actual coupling capacitor voltage and estimated coupling capacitor voltage are compared to verify the effectiveness of the v_{cy} observer. Figure 8.4 shows the sensed and estimated coupling capacitor voltage in α-β frame (v_{cy} and $\hat{v}_{cy}$) under inductive and capacitive load situation, which indicates that the observer can estimate the v_{cy} with fast dynamic response and low steady state error (maximum percentage error = 2%) under both inductive and capacitive loads.

After that, the simulation results of the proposed FS-MPC for TC*LC*-HAPF with typical and simplified model under different load conditions are given and discussed. Figure 8.5 shows the total harmonic distortion (THD) of phase *a* source current THDi_{sa} and settling time T_{st} with different firing angle α_x (i.e., different load conditions), in which the results by using the proposed simplified model with v_{cy} observer

Table 8.2 System parameters for simulation and experiment

	Parameters		Value
Grid parameters	V_{sx}, f_g	Simulation	11 kV, 50 Hz
		Experiment	110 V, 50 Hz
DC-link voltage	V_{DC}	Simulation	5 kV
		Experiment	60 V
TC*LC* parameters	L_c, L_{PF}, C_{PF}		5 mH, 30 mH, 160 μF
Sampling frequency	f_s		20 kHz
Non-linear load	L_{NL}, R_{NL}		30 mH, 45 Ω
Inductive linear load	L_{LL}, R_{LL}		60 mH, 15 Ω
Capacitive linear load	C_{LL}, R_{LL}		140 μF, 20 Ω
Observer gain	k_1, k_2		40,000, 6150

Fig. 8.4 Simulation results of sensed and estimated v_{cy} under **a** Nonlinear inductive load. **b** Nonlinear capacitive load

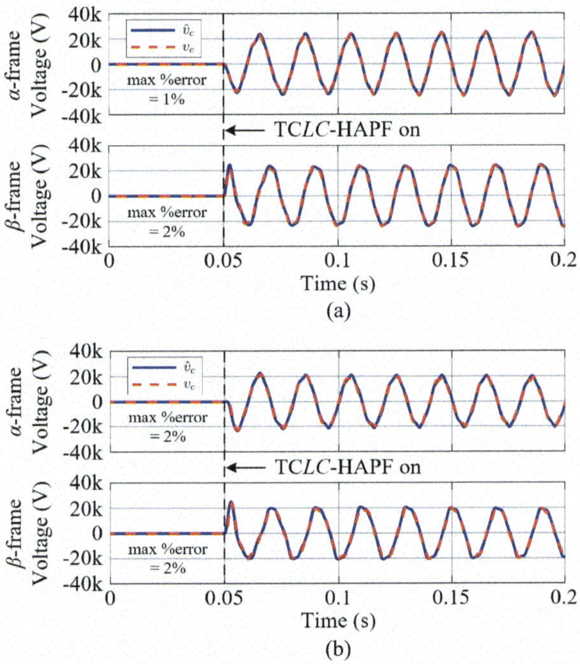

resemble the results with real model and simplified model with directly sensing v_{cy}. All the compensated THDi_{sa} are below 5% and the T_{st} are less than 7 ms under different firing angle value. From Figs. 8.4 and 8.5, the coupling capacitor voltage observer can be applied into the FS-MPC for the TC*LC*-HAPF in order to save the sensor cost while keeping a good compensation performance.

Additionally, the robustness is vital for the proposed FS-MPC with v_{cy} observer for TC*LC*-HAPF. To verify this, the simulation studies about the system parameters

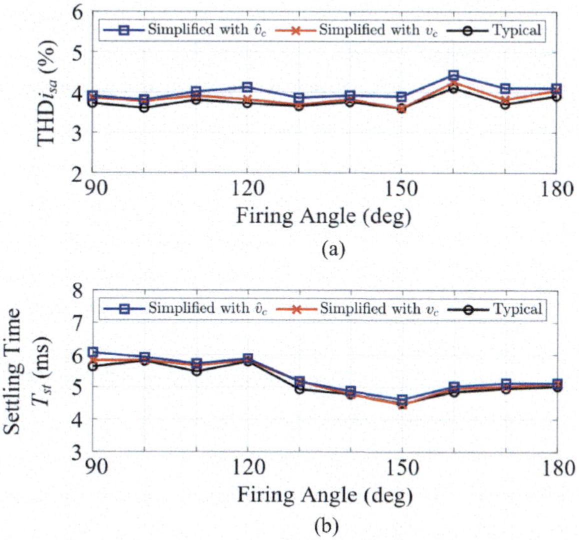

Fig. 8.5 Simulation results of FS-MPC with real and simplified model under different firing angles. **a** Phase a source current THDi_{sa}. **b** Settling time T_{st}

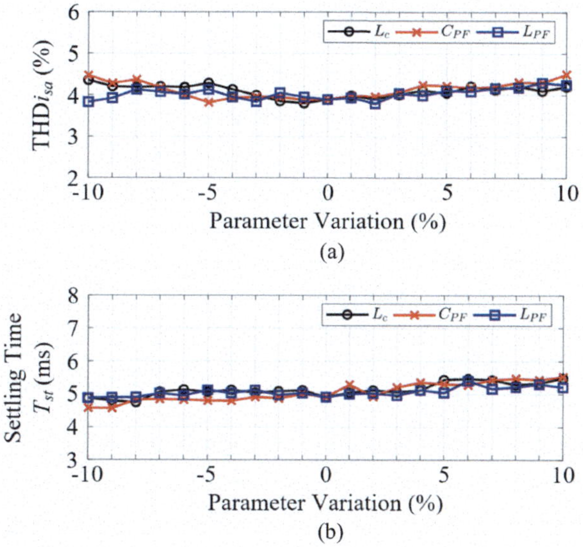

Fig. 8.6 Simulation results of FS-MPC with real model under system parameters variation. **a** Phase a source current THDi_{sa}. **b** Settling time T_{st}

variation are conducted. Figures. 8.6 and 8.7 show the results about the L_c, C_{PF} and L_{PF} vary from -10% to 10% of the nominal value with a step of 1%, in which the compensated THDi_{sa} and T_{st} of proposed simplified model with v_{cy} observer keep approximately the same level as that of the real model, respectively, (THDi_{sa} $< 5\%$ and $T_{st} < 6$ ms) during the variation of the system parameters. Therefore, the proposed FS-MPC with coupling capacitor voltage observer for the TCLC-HAPF can guarantee the compensation performance against the parameter variation condition.

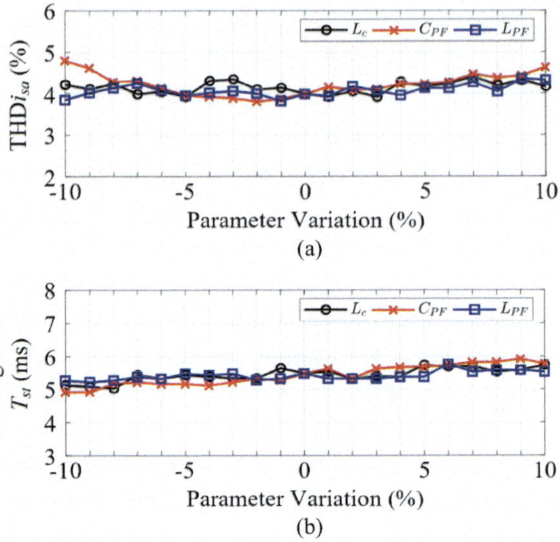

Fig. 8.7 Simulation results of FS-MPC with simplified model and v_{cy} observer under system parameters variation. **a** Phase a source current THDi_{sa}. **b** Settling time T_{st}

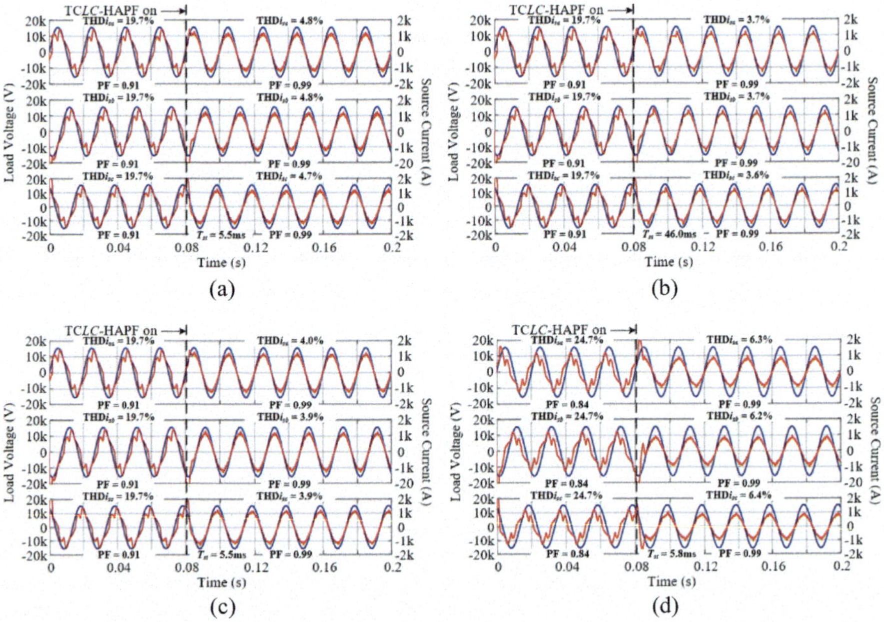

Fig. 8.8 Simulation waveforms of TC*LC*-HAPF dynamic compensation performance. **a** HCC under inductive loads; **b** MQPRC under inductive loads; **c** FS-MPC with v_{cy} observer under inductive loads; **d** FS-MPC with v_{cy} observer under capacitive loads

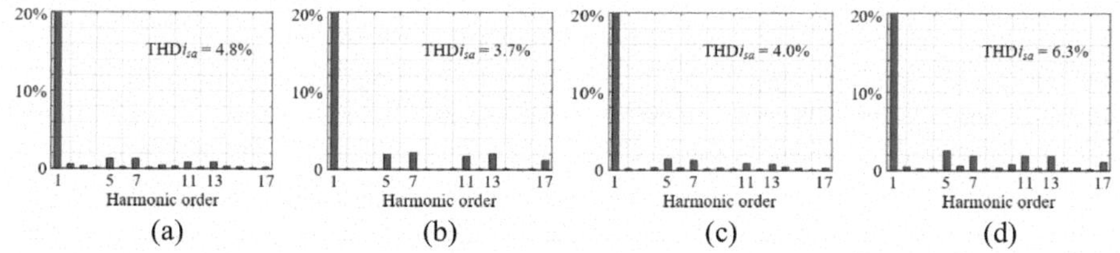

Fig. 8.9 Simulation results of i_{sa} harmonic spectrum after compensation. **a** HCC under inductive loads; **b** MQPRC under inductive loads; **c** FS-MPC with v_{cy} observer under inductive loads; **d** FS-MPC with v_{cy} observer under capacitive loads

Table 8.3 Simulation results before and after TC*LC*-HAPF compensation

		P_{sx} (kW)	Q_{sx} (kvar)	PF	i_{sx} (kA)	Worst THDi_{sx} (%)	T_{st} (ms)
Inductive Loads	Before Compensation	7904.9	3137.2	0.91	0.88	19.7	/
	HCC	7887.8	15.5	0.99	0.80	4.8	5.5
	MQPRC	7832.4	9.7	0.99	0.79	3.7	46.0
	FS-MPC	7853.5	10.8	0.99	0.80	4.0	5.5
Capacitive Loads	Before Compensation	5764.3	−3716.8	0.84	0.71	24.7	/
	HCC	5895.6	−16.2	0.99	0.60	7.3	5.8
	MQPRC	5865.1	−10.8	0.99	0.59	6.0	50.0
	FS-MPC	5880.0	−13.0	0.99	0.60	6.4	5.8

Finally, Figs. 8.8 and 8.9 shows the simulation waveforms and i_{sa} harmonic spectrum after compensation. From Figs. 8.8 and 8.9, the reactive power and current harmonics compensation capability of the proposed FS-MPC for the TC*LC*-HAPF has been well proved.

8.6 Experimental Results

Due to the limited conditions of the laboratory, a scaled down 110V–5kVA three-phase three-wire TC*LC*-HAPF experimental prototype was set up for the comparative experiments with different existing controllers of the TC*LC*-HAPF. Because the PIC in [12] is based on the synchronous reference frame (*d*-*q* frame), which is basically the same as the MQPRC in stationary reference frame (α-β frame) [13]. Therefore, the experimental results of PIC will not be given in this book. Figure 6.10 presents the experimental setup and Table 8.2 lists the system parameters for the experiment. The experimental results are summarized in Table 8.4 and presented in Figs. 8.10 and 8.11. Compared the experimental results with the simulation results, their differences are actually due to the equivalent series resistance (ESR) in hardware components, the resolution of the transducers, the digital computation error the response time of drivers for IGBTs and thyristors, and the noise in A/D signals and the controller. Finally, compared Table 8.4 with Table 8.3, their trends of the compensation performances among HCC, MQPRC and FS-MPC are consistent.

Regarding the IEEE standard 519-2014 [14], the acceptable total demand distortion (TDD) $\leq$ 15% with I_{sc}/I_L is 100 < 1000 scale (a small experimental prototype rating of 110V–5kVA). The nominal rate current is assumed to be equal to the fundamental load current for the worst-case analysis, which results in THD = TDD $\leq$ 15%. Therefore, this book evaluates the TC*LC*-HAPF current harmonics compensating performance by setting acceptable THD $\leq$ 15%.

(A) ***Hysteresis Current Control***: Figs. 8.10a, 8.11a and Table 8.4 show the inductive loads compensation performance of the HCC. The HCC provides a fast transient response with T_{st} = 8 ms when the TC*LC*-HAPF starts. And the steady-state THDi_{sx} is about 8.1%.

(B) ***Multi-Quasi-Proportional-Resonant Control***: Figs. 8.10b, 8.11b and Table 8.4 show the inductive loads compensation performance of the MQPRC with fundamental resonance gain scheduling. It takes about 70 ms to the steady-state after the TC*LC*-HAPF starts because of the slow response of integrators in the MQPRC. And the steady-state THDi_{sx} is about 6.5%.

(C) ***Finite-Set Model Predictive Control***: Figs. 8.10c, 8.11c and Table 8.4 show the inductive loads compensation performance of the proposed FS-MPC with v_{cy} observer under inductive loads. The proposal provides a very fast transient response with T_{st} = 8 ms when the TC*LC*-HAPF starts because of its directly optimal switching state selection and it is 8.75× faster than the MQPRC. The compensated THDi_{sx} is about 7% which is better than the HCC and similar as the MQPRC.

Table 8.4 Experimental results before and after TC*LC*-HAPF compensation

		P_{sx} (W)	Q_{sx} (var)	PF	i_{sx} (A)	Worst THDi_{sx} (%)	T_{st} (ms)
Inductive Loads	Before Compensation	680	400	0.84	7.3	21.5	/
	HCC	730	20	0.99	6.5	8.3	8
	MQPRC	720	10	0.99	6.4	6.7	70
	FS-MPC	730	10	0.99	6.4	7.0	8
Capacitive Loads	Before Compensation	770	−240	0.93	7.1	22.8	/
	HCC	800	−20	0.99	7.0	9.2	8
	MQPRC	800	−10	0.99	6.9	7.7	76
	FS-MPC	800	−10	0.99	6.9	8.2	8

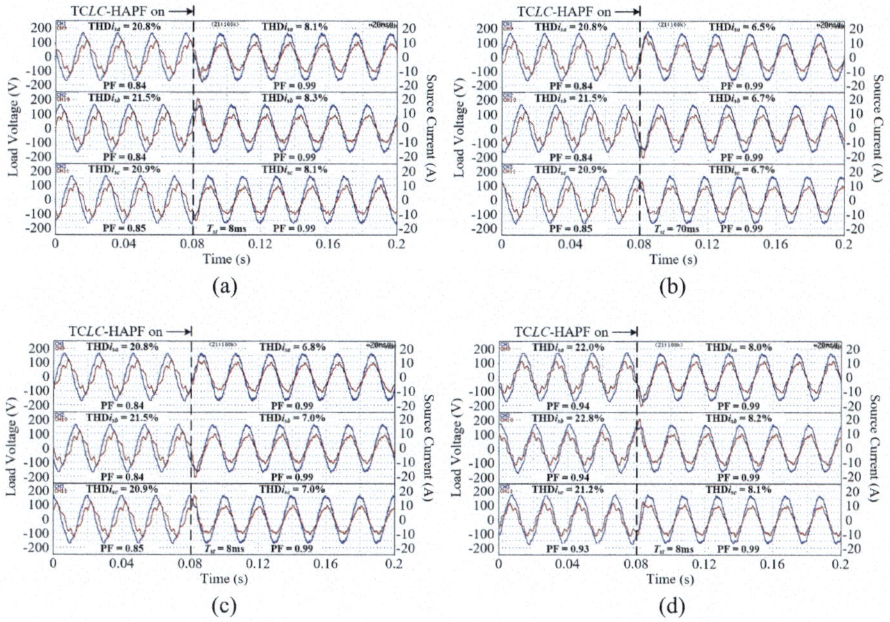

Fig. 8.10 Experimental waveforms of TC*LC*-HAPF dynamic compensation performance. **a** HCC under inductive loads; **b** MQPRC under inductive loads; **c** FS-MPC with v_{cy} observer under inductive loads; **d** FS-MPC with v_{cy} observer under capacitive loads

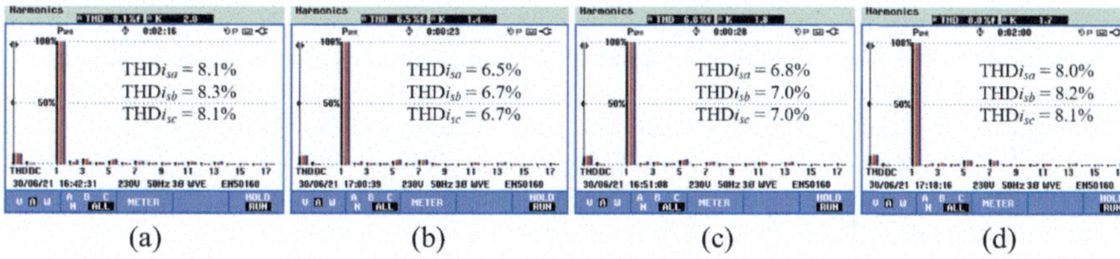

Fig. 8.11 Experimental results of i_{sx} harmonic spectrum after compensation. **a** HCC under inductive loads; **b** MQPRC under inductive loads; **c** FS-MPC with v_{cy} observer under inductive loads; **d** FS-MPC with v_{cy} observer under capacitive loads

Moreover, Figs. 8.10d, 8.11d and Table 8.4 show the capacitive loads compensation performance of the proposed FS-MPC with v_{cy} observer. The settling time is the same as the inductive loads case with $T_{st} = 8$ ms and the compensated THDi_{sx} can be reduced to 8% from 22%, which validate the effectiveness of the proposed FS-MPC even under capacitive loads case.

Table 8.4 also summarizes the experimental data with different control strategies in both capacitive and inductive loads, which evidently shows that the FS-MPC scheme performs a good compensation performance with fast transient response and low steady-state error.

8.7 Summary

This chapter proposes a FS-MPC with coupling capacitor voltage observer for the TC*LC*-HAPF, which can provide a fast transient response, low steady-state error and good robustness under both inductive and capacitive load conditions. The real predictive control model is first constructed to realize the FS-MPC for the TC*LC*-HAPF. And the simplified model is proposed afterward for the ease of implementing the FS-MPC on the TC*LC*-HAPF. To reduce the hardware costs caused by the coupling capacitor voltage sensing circuits, a coupling capacitor voltage observer is designed. Also, the delay compensation is considered for the actual implementation. The effectiveness and performance of the proposal are verified by simulated and experimental results. Among the existing control schemes for the TC*LC*-HAPF, the proposed model predictive controlled TC*LC*-HAPF obtains a fast transient response speed similar as the HCC and good steady-state performance similar as the MQPRC, respectively.

References

1. C.-S. Lam, L. Wang, S.-I. Ho, M.-C. Wong, Adaptive thyristor-controlled LC-hybrid active power filter for reactive power and current harmonics compensation with switching loss reduction. IEEE Trans. Power Electron. **32**(10), 7577–7590 (2017)
2. L. Wang, C.-S. Lam, M.-C. Wong, Selective compensation of distortion, unbalanced and reactive power of a thyristor-controlled LC-coupling hybrid active power filter (TCLC-HAPF). IEEE Trans. Power Electron. **32**(12), 9065–9077 (2017)
3. W.-K. Sou, C.-W. Chao, C. Gong, C.-S. Lam, C.-K. Wong, Analysis, design and implementation of multi-quasi-proportional-resonant controller for thyristor-controlled LC-coupling hybrid active power filter (TCLC-HAPF). IEEE Trans. Ind. Electron. **69**(1), 29–40 (2022)
4. L. Wang, C.-S. Lam, M.-C. Wong, Modeling and parameter design of thyristor-controlled LC-coupled hybrid active power filter (TCLC-HAPF) for unbalanced compensation. IEEE Trans. Ind. Electron. **64**(3), 1827–1840 (2017)
5. C. Zheng, T. Dragičević, F. Blaabjerg, Current-sensorless finite-set model predictive control for LC-filtered voltage source inverters. IEEE Trans. Power Electron. **35**(1), 1086–1095 (2020)
6. J. Kukkola, M. Hinkkanen, K. Zenger, Observer-based state-space current controller for a grid converter equipped with an LCL filter: analytical method for direct discrete-time design. IEEE Trans. Ind. Appl. **51**(5), 4079–4090 (2015)
7. S. Buso, P. Mattavelli, *Digital control in power electronics* (Morgan & Claypool, 2015)
8. D. Xiao, K.S. Alam, M. Norambuena, M.F. Rahman, J. Rodriguez, Modified modulated model predictive control strategy for a grid-connected converter. IEEE Trans. Ind. Electron. **68**(1), 575–585 (2021)
9. C. Zheng, T. Dragičević, Z. Zhang, J. Rodriguez, F. Blaabjerg, Model predictive control of LC-filtered voltage source inverters with optimal switching sequence. IEEE Trans. Power Electron. **36**(3), 3422–3436 (2021)
10. P. Falkowski, A. Sikorski, Finite control set model predictive control for grid-connected AC–DC converters with LCL filter. IEEE Trans. Ind. Electron. **65**(4), 2844–2852 (2018)
11. H. Akagi, Y. Kanazawa, A. Nabae, Instantaneous reactive power compensators comprising switching devices without energy storage components. IEEE Trans. Ind. Appl. **IA-20**(3), 625–630 (May 1984)

12. S. Rahmani, A. Hamadi, K. Al-Haddad, L.A. Dessaint, A combination of shunt hybrid power filter and thyristor-controlled reactor for power quality. IEEE Trans. Ind. Electron. **61**(5), 2152–2164 (May 2014)
13. C. Lascu, L. Asiminoaei, I. Boldea, F. Blaabjerg, Frequency response analysis of current controllers for selective harmonic compensation in active power filters. IEEE Trans. Ind. Electron. **56**(2), 337–347 (2009)
14. IEEE recommended practices and requirements for harmonic control in electrical power systems (2014). IEEE Standard 519-2014

Chapter 9
Finite-Set Hybrid Model Predictive Control for TC*LC*-HAPF

9.1 Introduction

This chapter proposes a finite-set hybrid model predictive control (FS-HMPC) by merging the sliding mode control (SMC) and finite-set model predictive control (FS-MPC). Compared to FS-MPC, the proposed control scheme can further improve the steady-state performance and robustness of TC*LC*-HAPF. The sliding coefficient design, stability and robustness analysis are also presented. Finally, simulation and experiment studies are given to verify the proposal.

The contribution of this chapter is summarized below:

(1) Propose the FS-HMPC that the predictive model is included into a sliding surface function for the rolling optimization;
(2) Provide the stability and robustness analysis for the proposal;
(3) Derive the design procedure of the predictive sliding surface function;
(4) Verify the effectiveness and performance of the proposal by simulation and experimental results.

9.2 Modeling of TC*LC*-HAPF

Since the coupling capacitor voltage v_{cy} equals to the voltage of the thyristor-controlled reactor, a simplified model of the TC*LC*-HAPF was proposed for implementing the FS-MPC in Chap. 8. In this chapter, the simplified model is used again for the ease of implementing FS-HMPC. According to Fig. 8.1b, the simplified system dynamic model of TC*LC*-HAPF is derived as (9.1).

$$\begin{cases} L_c \frac{di_{cy}}{dt} = v_{invy} - v_y - v_{cy} \\ C_{PF} \frac{dv_{cy}}{dt} = i_{cy} \end{cases} \tag{9.1}$$

C.-S. Lam et al., *Advanced Controllers Design for Hybrid Active Power Filters*,
https://doi.org/10.1007/978-981-97-6847-9_9

9.3 Design of Hybrid Finite-Set Model Predictive Control for TCLC-HAPF

9.3.1 Fundamentals of FS-MPC and Hysteresis SMC

(A) ***FS-MPC***: The FS-MPC utilizes a discrete predictive model to forecast the future state variables and applies optimal switching state for the switching devices to fulfill the control objectives, in which the optimal switching state is determined by finding the minimum value of a cost function using different system input.

Based on (9.1), the discrete predictive model can be formulated as (9.2) with the help of Euler method, where T_s is the sampling period. A_d and B_d are coefficient matrices. x, u are state variable and system input.

$$\underbrace{\begin{bmatrix} i_{cy}(k+1) \\ v_{cy}(k+1) \end{bmatrix}}_{x(k+1)} = \underbrace{\begin{bmatrix} 1 & -\frac{T_s}{L_c} \\ \frac{T_s}{C_{PF}} & 1 \end{bmatrix}}_{A_d} \underbrace{\begin{bmatrix} i_{cy}(k) \\ v_{cy}(k) \end{bmatrix}}_{x(k)} + \underbrace{\begin{bmatrix} \frac{T_s}{L_c} \\ 0 \end{bmatrix}}_{B_d} \underbrace{\begin{bmatrix} v_{invy}(k) - v_y(k) \end{bmatrix}}_{u(k)} \tag{9.2}$$

The control objective of the FS-MPC for TCLC-HAPF is to minimize the error of compensating current i_c, thus minimizing the compensated current total harmonic distortion (THDi) and reactive power, hence the cost function is defined as

$$J = \left(i_{c\alpha}{}^*(k+1) - i_{c\alpha}(k+1)\right)^2 + \left(i_{c\beta}{}^*(k+1) - i_{c\beta}(k+1)\right)^2 \tag{9.3}$$

where the notation '*' denotes the reference value. $i_{c\alpha}$ and $i_{c\beta}$ are the compensating current in α and β frame, respectively. The $i_{cy}{}^*(k+1)$ can be calculated by using Lagrange extrapolation method [1] with $i_{cy}{}^*(k)$ obtained from (8.22).

By enumerating 8 possible inverter voltage vectors $v_{invy}(k)$ in (9.2), the optimal one that can generate the optimal $i_{cy}(k+1)$ to minimize (9.3) will be selected and then applied to the active inverter part of the TCLC-HAPF. The relationship between the switching states of each phase S_x and inverter voltage vectors v_{invy} is tabulated in Table 8.1. Also, S_x is given in (8.21).

(B) ***Hysteresis SMC***: The SMC forces the system trajectory converges to a sliding surface and keeps sliding on it. Same with the FS-MPC, the control objective is to guarantee the compensating current i_c equals to its reference value. The sliding surface function s_x is defined as [2]

$$s_x = \lambda\sigma_{1x} + \sigma_{2x} \tag{9.4}$$

where λ is the sliding coefficient. The TCLC-HAPF system can reach the sliding surface if and only if the sliding mode existence condition $s_x\dot{s}_x < 0$ is satisfied, and this condition can only be achieved by designing a suitable λ value [1]. σ_{1x} and σ_{2x} are sliding variables of each phase. And they are shown in below [3].

$$\begin{cases} \sigma_{1x} = i_{cx}{}^{*} - i_{cx} \\ \sigma_{2x} = \frac{d\sigma_{1x}}{dt} \end{cases} \tag{9.5}$$

Based on (8.21) and (9.4), the inverter trigger signals T_{1x} and T_{2x} can be obtained by following piecewise function.

$$S_x = \begin{cases} 1, & s_x > 0 \\ 0, & s_x < 0 \end{cases} \tag{9.6}$$

9.3.2 *Predictive Sliding Surface Function*

The concept of the proposed FS-HMPC is to execute the rolling optimization of predictive sliding surface function that is formed by the predictive model and sliding surface function, thence enhancing the steady-state performance and robustness of the TC*LC*-HAPF compared to the FS-MPC.

Substitute (9.5) into (9.4), the sliding surface function in complex frame α and β becomes

$$s_y = \lambda(i_{cx}{}^{*} - i_{cy}) + \frac{d(i_{cx}{}^{*} - i_{cy})}{dt} \tag{9.7}$$

and its discretized form is shown in (9.8), where subscript '*d*' represents the derivative term. The derivative $i_{cy,d}$ is substituted by the second order derivative of v_{cy} based on (9.1) to avoid distortion caused by high-frequency switching noise in i_{cy}. And the derivative $i_{cy,}{}^{*}{}_{d}$ is formulated based on Euler method.

$$\begin{cases} s_y(k) = \lambda[i_{cy}{}^{*}(k) - i_{cy}(k)] + i_{cy,d}{}^{*}(k) - i_{cy,d}(k) \\ i_{cy,d}{}^{*}(k) = \frac{i_{cy}{}^{*}(k) - i_{cy}{}^{*}(k-1)}{T_s} \\ i_{cy,d}(k) = \frac{C_{PF}}{T_s^2}\left[v_{cy}(k) - 2v_{cy}(k-1) + v_{cy}(k-2)\right] \end{cases} \tag{9.8}$$

Therefore, the predictive sliding surface function is defined as (9.9) based on predictive model (9.2) and sliding surface function (9.8). Note that $i_{cy}(k+1)$ and $v_{cy}(k+1)$ are obtained from (9.2), $v_{cy}(k)$ can either be directly measured by sensing circuit or estimated by observer. In this work, the observer is used to lower the hardware cost and the design can refer to Sect. 8.3.2.

$$\begin{aligned} s_y(k+1) = {} & \lambda\left[i_{cy}^{*}(k+1) - i_{cy}(k+1)\right] + \frac{i_{cy}^{*}(k+1) - i_{cy}^{*}(k)}{T_s} \\ & - \frac{C_{PF}}{T_s^2}\left[v_{cy}(k+1) - 2v_{cy}(k) + v_{cy}(k-1)\right] \end{aligned} \tag{9.9}$$

9.3.3 Cost Function Design

The cost function design is based on the characteristic of the SMC and the control objective of the FS-MPC. The SMC drives the sliding surface function to the ideal sliding surface, i.e. $s = 0$. And the FS-MPC hopes to minimize the error through the predictive model. Hence, the cost function of the proposed FS-HMPC can be set as finding out which predictive sliding surface is closed to 0. Mathematically, that is

$$J_s = (0 - s_\alpha(k+1))^2 + (0 - s_\beta(k+1))^2 \tag{9.10}$$

Finally, by enumerating the possible inverter voltage vectors $v_{invy}(k)$ shown in Table 8.1, the optimal inverter voltage that can generate the optimal $i_{cy}(k+1)$ and $s_y(k+1)$ to minimize (9.10) will be selected and applied its corresponding trigger signals to the active inverter part of the TC*LC*-HAPF.

9.3.4 Stability and Robustness Analysis

(A) ***Stability Analysis***: To analyze the stability issue, the famous Lyapunov stability analysis is usually applied [4–6]. But this book studies the stability problem of FS-HMPC based on the discrete SMC theory [7] in order to provide the analysis in an intuitive way. The purpose of the stability analysis section is to prove that the proposed controlled TC*LC*-HAPF is a bounded input, bounded output (BIBO) stable system. According to [6, 7], the optimal inverter voltage vector $v_{invy,op}$ applied to the active inverter part is defined as

$$v_{invy,op}(k) = \overline{v}_{invy}(k) + \delta_y(k) \tag{9.11}$$

where $\overline{v}_{invy}$ is the continuous inverter voltage vector generated by the FS-HMPC and δ_y is the quantization error. The $v_{invy,op}$ is bounded in the finite set with bounded DC-link voltage (as shown in Table 8.1), hence the δ_y is also bounded with $|\delta_y| \leq \varepsilon$, where constant $\varepsilon > 0$. The $\overline{v}_{invy}$, which is shown in (9.12), can be obtained by setting (9.9) equals to 0 (i.e. ideal sliding surface $s = 0$ is reached) with the help of (9.2).

$$\begin{aligned} \overline{v}_{invy}(k) = \Big\{ & \lambda L_c\big[i_{cy}{}^*(k+1) - i_{cy}(k)\big] + \lambda T_s\big[v_{cy}(k) + v_y(k)\big] \\ & + L_c i_{cy,d}{}^*(k+1) - \tfrac{L_c}{T_s} i_{cy}(k) + v_{cy}(k) + v_y(k) \\ & + \tfrac{L_c C_{PF}}{T_s^2}\big[v_{cy}(k) - v_{cy}(k-1)\big]\Big\} / (\lambda T_s + 1) \end{aligned} \tag{9.12}$$

Then, take into account the parameter mismatch in the predictive model (9.2), hence the predicted state variables ($i_{cy}(k+1)$ and $v_{cy}(k+1)$) involve with the model mismatch constant m_{1y} and m_{2y} as shown in (9.13). They are assumed to be bounded with $|m_{1y}| \le M_1$ and $|m_{2y}| \le M_2$, where $M_1 > 0$ and $M_2 > 0$.

$$\underbrace{\begin{bmatrix} i_{cy}(k+1) \\ v_{cy}(k+1) \end{bmatrix}}_{\mathrm{x}(k+1)} = \underbrace{\begin{bmatrix} 1 & -\frac{T_s}{L_c} \\ \frac{T_s}{C_{PF}} & 1 \end{bmatrix}}_{\mathrm{A_d}} \underbrace{\begin{bmatrix} i_{cy}(k) \\ v_{cy}(k) \end{bmatrix}}_{\mathrm{x}(k)} + \underbrace{\begin{bmatrix} \frac{T_s}{L_c} \\ 0 \end{bmatrix}}_{\mathrm{B_d}} \underbrace{\left[v_{invy}(k) - v_y(k)\right]}_{\mathrm{u}(k)} + \begin{bmatrix} m_{1y}(k) \\ m_{2y}(k) \end{bmatrix} \tag{9.13}$$

Substitute (9.12) and (9.13) into (9.9), the dynamics of predictive sliding surface becomes

$$s_y(k+1) = -[\frac{\lambda T_s + 1}{L_c}\delta_y(k) + \frac{\lambda T_s + 1}{T_s}m_{1y}(k) + \frac{C_{PF}}{T_s^2}m_{2y}(k)] \tag{9.14}$$

Finally, considering the boundary value of δ_y, m_{1y} and m_{2y}, the discrete SMC band H can be obtained as

$$H = \left|s_y(k+1)\right| \le \frac{\lambda T_s + 1}{L_c}\varepsilon + \frac{\lambda T_s + 1}{T_s}M_1 + \frac{C_{PF}}{T_s^2}M_2 \tag{9.15}$$

(9.15) indicates that the $s_y(k + 1)$ of the proposed FS-HMPC is ultimately bounded, which means the output compensating current i_{cy} of the proposed controlled TC*LC*-HAPF is also bounded. Therefore, it is a BIBO stable system [8].

(B) ***Robustness Analysis***: Since the robustness is the ability of a system to deal with uncertainty, the assumptions of the stability analysis can be reused. For the proposed FS-HMPC, since the control objective is tracking $s_y(k + 1) = 0$, by dividing λ on both sides of (9.9), $s_y(k + 1)$ will become (9.16).

$$\begin{aligned} s_y(k+1) = {} & [i_{cy}{}^*(k+1) - i_{cy}(k+1)] + \frac{i_{cy}{}^*(k+1) - i_{cy}{}^*(k)}{\lambda T_s} \\ & - \frac{C_{PF}}{\lambda T_s^2}[v_{cy}(k+1) - 2v_{cy}(k) + v_{cy}(k-1)] \end{aligned} \tag{9.16}$$

From (9.16), when $\lambda \to \infty$, the 2nd and 3rd terms will approach 0. In this case, substituting $s_y(k + 1)$ to cost function (9.10), the cost function of the proposed FS-HMPC will become (9.17), which equals to the cost function of the conventional FS-MPC (9.3). Hence, the proposed FS-HMPC degrades to the conventional FS-MPC when $\lambda \to \infty$. In addition, from the deduced discrete SMC band formula (9.15), the band H is proportional to the sliding coefficient λ in the steady-state. From (9.15), because the λ of conventional FS-MPC $\to \infty$, the FS-MPC would produce a larger H than the proposed FS-HMPC. From the discrete SMC [7, 9, 10], the larger H will result in poorer robustness and steady-state performance. Consequently, the proposed FS-HMPC will have a better robustness and steady-state performance because of the smaller H.

$$J_s = (i_{c\alpha}(k+1) - i_{c\alpha}{}^*(k+1))^2 + (i_{c\beta}(k+1) - i_{c\beta}{}^*(k+1))^2 \tag{9.17}$$

It is worth mentioning that the above analysis can explain the strong robustness of both FS-MPC and FS-HMPC under the coupling capacitor voltage v_c mismatch (or observation error of v_c observer). It is because from (5.15), the value of v_{cy} mismatch constant M_2 is not related to λ. Even if there is an error in v_c, the M_2 will not be amplified by the λ and the H will not be further increased, which means both control scheme can still maintain a good steady-state performance.

9.3.5 Sliding Coefficient Design

In (9.9), there is a sliding coefficient λ needs to be designed. In general, this coefficient is designed based on the trial-and-error method [2]. However, it determines the position of the closed-loop poles for the TCLC-HAPF system. Therefore, the poles placement method [2, 7] is used in this work to simplify the tuning workload of λ.

For getting the closed-loop state-space equation, substitute the optimal inverter voltage vector $v_{invy,op}$ (9.12) into the predictive model (9.2), which yields

$$\mathrm{x}(k+1) = (\mathrm{A_d} + \mathrm{B_d}\mathrm{E})\mathrm{x}(k) + \mathrm{F}(k) \tag{9.18}$$

where $\mathbf{A_d}$ and $\mathbf{B_d}$ are the coefficient matrices that are same as (9.2), $\mathbf{E} = \left[-L_c/T_s \;\; 1 - L_c C_{PF}/T_s^2(\lambda T_s + 1)\right]$ and $\mathbf{F}(k)$ is the other part that is not related to state variables $\mathbf{x}(k)$. From that, the characteristic equation is then obtained as below, and $\mathbf{I}$ is a 2×2 identity matrix.

$$\det[z\mathrm{I} - (\mathrm{A_d} + \mathrm{B_d}\mathrm{E})] = 0 \tag{9.19}$$

After that, the closed-loop poles of (9.18) can be calculated as

$$z_{1,2} = \frac{1}{2} \pm \frac{\sqrt{(\lambda T_s - 3)(\lambda T_s + 1)}}{2(\lambda T_s + 1)} \tag{9.20}$$

From (9.20), it is easy to observe that two closed-loop poles will meet each other when $\lambda T_s - 3 = 0$, which means $\lambda = 3/T_s$. In this case, $z_1 = z_2 = 0.5 < 1$ can guarantee the closed-loop stability and get a balance between dynamic performance and oscillation rejection. Therefore, λ is set as $3/T_s$ in this chapter.

9.4 Overall Control Block Diagram

The overall control block diagram of the proposed control scheme is illustrated in Fig. 9.1, which can be separated into two parts: TC*LC* part control and active inverter part control. Sections. 6.4 and 8.4 have already presented the TC*LC* part control and the reference compensating current calculation of active inverter part control, respectively. To avoid repetition, they will not be covered again in this chapter.

By sending the reference compensating current to the proposed FS-HMPC, the proposed control will select the switching state S_x and generate the corresponding inverter trigger signals T_{1x} and T_{2x} of each phase.

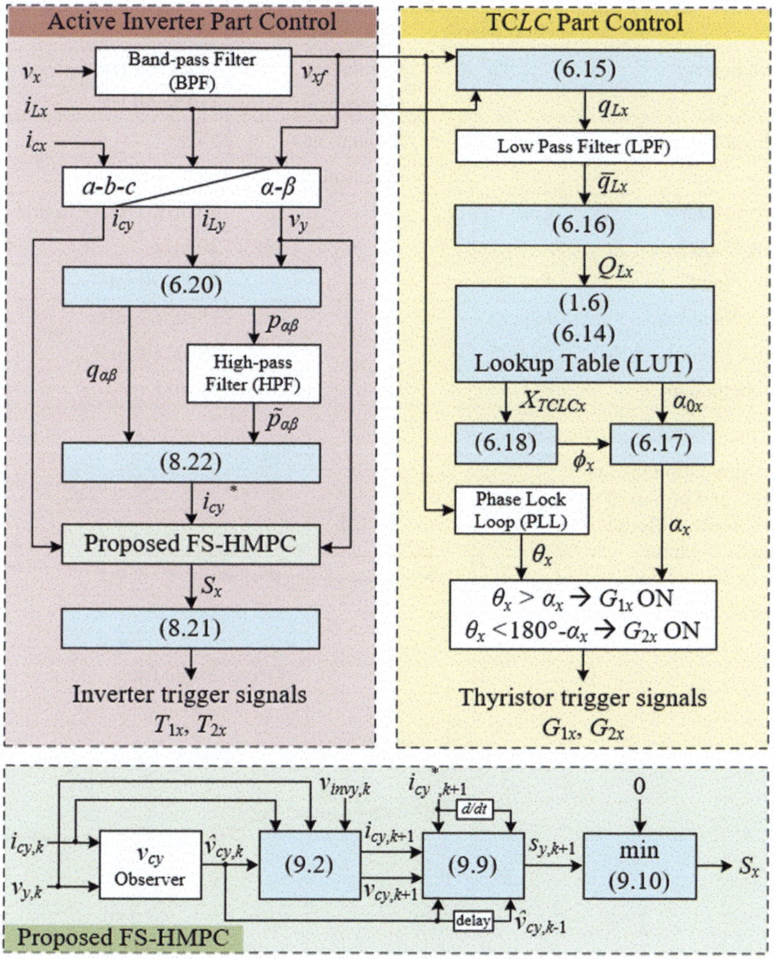

Fig. 9.1 Overall control block diagram of TC*LC*-HAPF for FS-MPC

9.5 Simulation Results

For verifying the performance of the proposed FS-HMPC for the TC*LC*-HAPF under the 11 kV application, a simulation study using MATLAB/Simulink is provided in this section. The system parameters for simulation are listed in Table 9.1. A DC voltage source is adopted for providing the DC-link voltage in order to ease the verification. The simulation results are shown in Figs. 9.2, 9.3, 9.4 and 9.5 and Tables 9.2, 9.3 and 9.4. Note that all the transient response (settling time T_{st}) of this work is obtained after connecting the TC*LC*-HAPF to the system with load.

Table 9.1 System parameters for simulation and experiment

	Parameters		Value
Grid parameters	V_{gx}, f_g	Simulation	11 kV, 50 Hz
		Experiment	110 V, 50 Hz
DC-link voltage	V_{dc}	Simulation	5 kV
		Experiment	60 V
TC*LC* parameters	L_{PF}, C_{PF}, L_c		30 mH, 160 μF, 5 mH
Sampling frequency	f_s		20 kHz
Non-linear load	R_{NL}, L_{NL}		45 Ω, 30 mH
Inductive linear load	R_{LL}, L_{LL}		15 Ω, 60 mH
Capacitive linear load	R_{LL}, C_{LL}		20 Ω, 140 μF
Sliding coefficient	λ		60,000

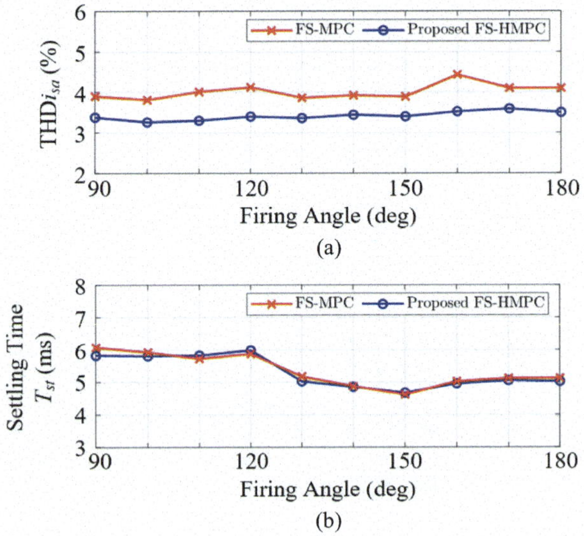

Fig. 9.2 Simulation results of FS-MPC and proposed FS-HMPC under different firing angles. **a** Phase *a* source current THDi_{sa}. **b** Settling time T_{st}

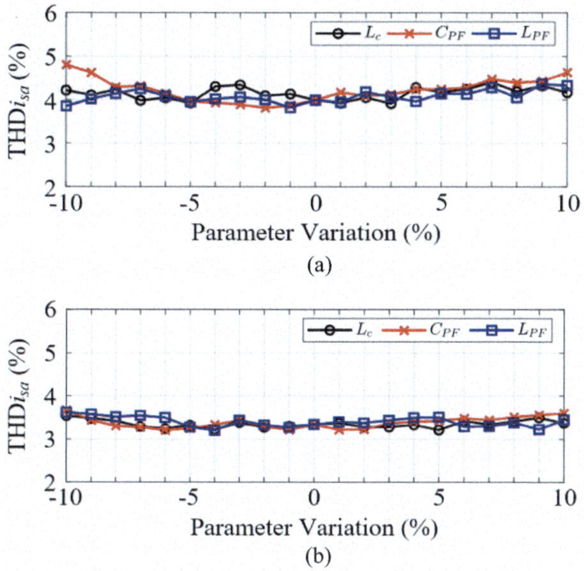

Fig. 9.3 Simulated phase *a* source current THDi_{sa} of FS-HMPC and FS-MPC under system parameter variation. **a** FS-MPC. **b** Proposed FS-HMPC

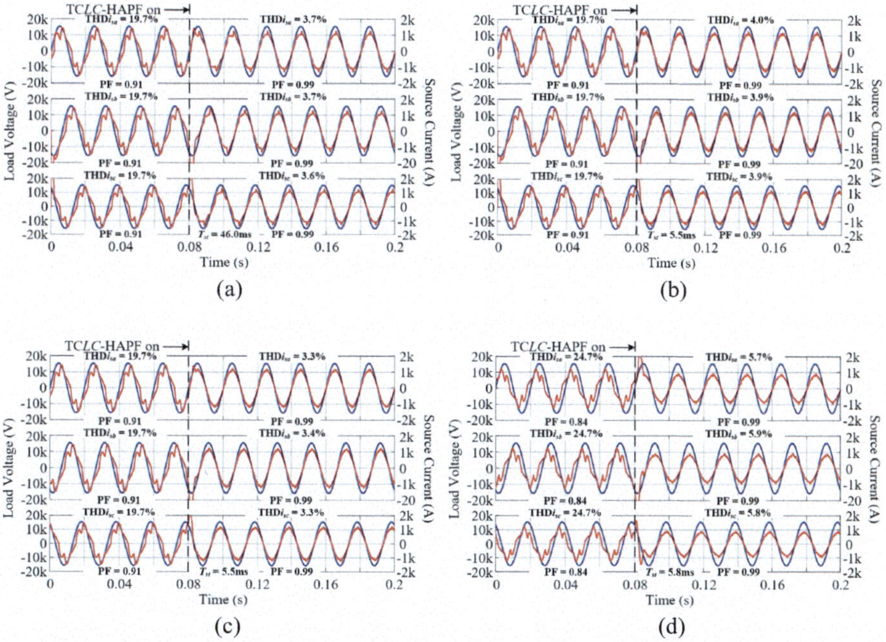

Fig. 9.4 Simulation waveforms of TC*LC*-HAPF dynamic compensation performance. **a** MQPRC under inductive loads; **b** FS-MPC under inductive loads; **c** FS-HMPC under inductive loads; **d** FS-HMPC under capacitive loads

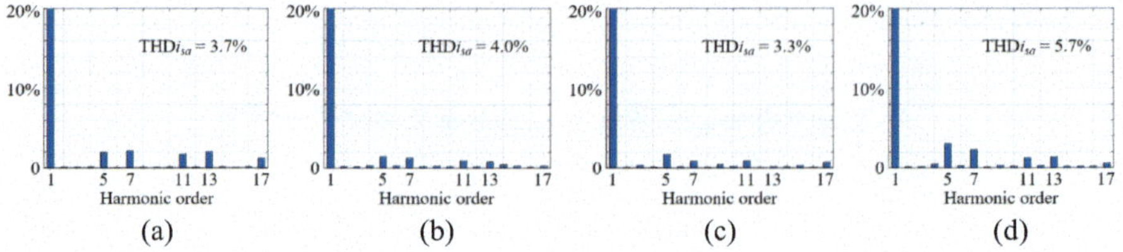

Fig. 9.5 Simulation results of i_{sa} harmonic spectrum after compensation. **a** MQPRC under inductive loads; **b** FS-MPC under inductive loads; **c** FS-HMPC under inductive loads; **d** FS-HMPC under capacitive loads

Table 9.2 Mean value of THD change under ±10% parameter variation

	Mean Value of THD Change (%)		
	L_c	C_{PF}	L_{PF}
FS-MPC	0.18	0.30	0.13
Proposed FS-HMPC	0.08	0.12	0.11

Table 9.3 Average switching frequency and efficiency comparison of FS-MPC and proposed FS-HMPC

		Simulation		Experiment	
		Average Switching Frequency (Hz)	Efficiency (%)	Average Switching Frequency (Hz)	Efficiency (%)
Inductive Loads	FS-MPC	2980	98.7	1830	97.0
	FS-HMPC	3200	98.6	2100	96.9
Capacitive Loads	FS-MPC	2380	97.6	1520	93.4
	FS-HMPC	2500	97.5	1650	93.2

Table 9.4 Simulation results before and after TCLC-HAPF compensation

		P_{sx} (kW)	Q_{sx} (kvar)	PF	i_{sx} (kA)	Worst THDi_{sx} (%)	T_{st} (ms)
Inductive Loads	Before Compensation	7904.9	3137.2	0.91	0.88	19.7	/
	MQPRC	7832.4	9.7	0.99	0.79	3.7	46.0
	FS-MPC	7853.5	10.8	0.99	0.80	4.0	5.5
	FS-HMPC	7848.2	10.2	0.99	0.80	3.4	5.5
Capacitive Loads	Before Compensation	5764.3	−3716.8	0.84	0.71	24.7	/
	MQPRC	5865.1	−10.8	0.99	0.59	6.0	50.0
	FS-MPC	5880.0	−13.0	0.99	0.60	6.4	5.8
	FS-HMPC	5875.7	−12.5	0.99	0.60	5.9	5.8

The simulation results of the proposed FS-HMPC for TC*LC*-HAPF under different load conditions are given in Fig. 9.2. It shows the total harmonic distortion (THD) of compensated source current (phase a) and settling time T_{st} against different TC*LC* part's firing angle α_x (i.e. different loads). The results tell that the THDi_{sa} of the proposed FS-HMPC is lower than the FS-MPC, which means a better steady-state performance. And the T_{st} of the proposed FS-HMPC is similar to the FS-MPC because both of them are based on the optimal voltage vector selection and do not need any modulation.

The simulation results under the system parameters variation are given to verify the robustness of the proposed FS-HMPC. Figure 9.3 presented the THDi_{sa} compensated by the FS-HMPC and FS-MPC under L_c, C_{PF} and L_{PF} vary in $\pm 10\%$ of the nominal value with step size of 1%. It clearly shows that the compensated THDi_{sa} of the proposed FS-HMPC is more convergent than the FS-MPC. To better evaluate the results, a mean value of THD change is used in this work, which is calculated by

$$Mean = \frac{\sum_{n=-10}^{10} \left| THDi_{sa,0} - THDi_{sa,n} \right|}{20} \tag{9.21}$$

where THD$i_{sa,0}$ is the THDi_{sa} under nominal value, and THD$i_{sa,n}$ is the THDi_{sa} under $n\%$ of parameter variation. The corresponding result is summarized in Table 9.2. It shows that the mean value of THD change of the FS-HMPC under all three parameters variation independently is smaller than that of the FS-MPC. In short, the robustness of the proposed FS-HMPC is improved.

The simulated average switching frequency and efficiency comparison of FS-MPC and proposed FS-HMPC is given in Table 9.3. The average switching frequency is obtained by counting the rising edge number of phase a IGBT's trigger signal. And the efficiency is obtained by calculating the ratio of the load active power to the source active power. From the results, the average switching frequency of the proposed FS-HMPC is higher than that of FS-MPC, which ensures a better steady-state performance. And the efficiency of the proposed FS-HMPC is comparable to that of FS-MPC even though the average switching frequency is higher.

Finally, Figs. 9.4 and 9.5 shows the simulation waveforms and i_{sa} harmonic spectrum after compensation. From Figs. 9.4 and 9.5, the reactive power and current harmonics compensation capability of the proposed FS-HMPC for the TC*LC*-HAPF has been well verified.

9.6 Experimental Results

Comparative experiments are conducted by using a scaled down 110V–5kVA three-phase three-wire TC*LC*-HAPF due to the limitation of laboratory. The experimental setup is presented in Fig. 6.10, and the parameters are listed in Table 9.1. The experimental results are summarized in Figs. 9.6, 9.7, Tables 9.3, 9.5 and Table 9.6. Compared the experimental results with the simulation results, their differences are actually due to the equivalent series resistance (ESR) in hardware components, the resolution of the transducers, the digital computation error the response time of drivers for IGBTs and thyristors, and the noise in A/D signals and the controller.

Regarding the IEEE standard 519-2014 [11], the acceptable total demand distortion (TDD) $\leq 15\%$ with I_{sc}/I_L is 100 < 1000 scale (a small experimental prototype rating of 110V–5kVA). The nominal rate current is assumed to be equal to the fundamental load current for the worst-case analysis, which results in THD = TDD $\leq 15\%$. Therefore, this book evaluates the TC*LC*-HAPF current harmonics compensating performance by setting acceptable THD $\leq 15\%$.

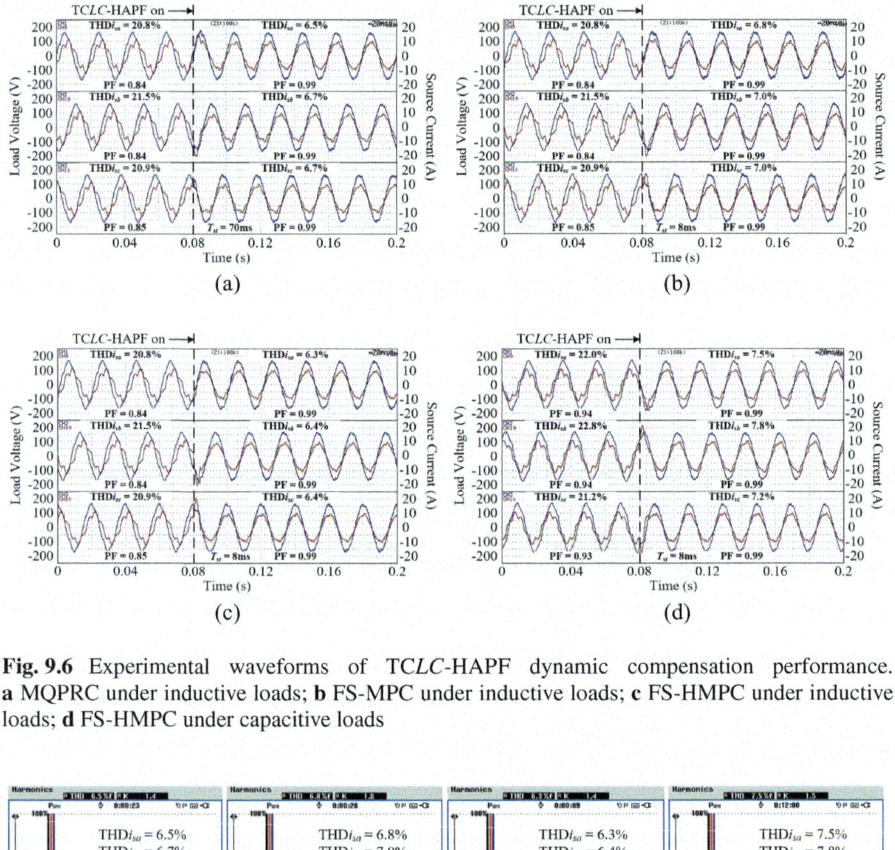

Fig. 9.6 Experimental waveforms of TC*LC*-HAPF dynamic compensation performance. **a** MQPRC under inductive loads; **b** FS-MPC under inductive loads; **c** FS-HMPC under inductive loads; **d** FS-HMPC under capacitive loads

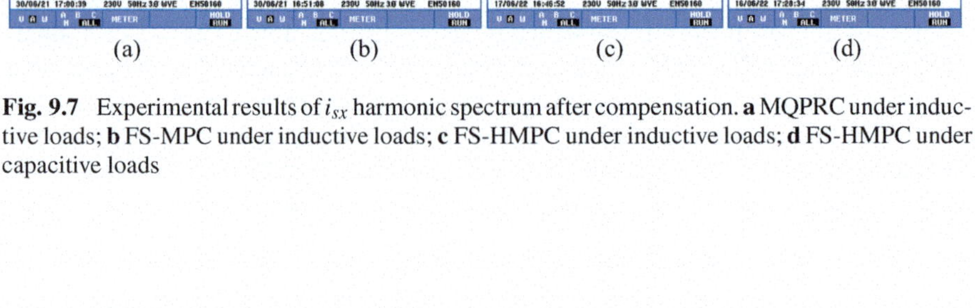

Fig. 9.7 Experimental results of i_{sx} harmonic spectrum after compensation. **a** MQPRC under inductive loads; **b** FS-MPC under inductive loads; **c** FS-HMPC under inductive loads; **d** FS-HMPC under capacitive loads

Table 9.5 Experiment results before and after TC*LC*-HAPF compensation

		P_{sx} (W)	Q_{sx} (var)	PF	i_{sx} (A)	Worst THDi_{sx} (%)	T_{st} (ms)
Inductive Loads	Before Compensation	680	400	0.84	7.3	21.5	/
	MQPRC	720	10	0.99	6.4	6.7	70
	FS-MPC	730	10	0.99	6.4	7.0	8
	FS-HMPC	730	10	0.99	6.4	6.4	8
Capacitive Loads	Before Compensation	770	−240	0.93	7.1	22.8	/
	MQPRC	800	−10	0.99	6.9	7.7	76
	FS-MPC	800	−10	0.99	6.9	8.2	8
	FS-HMPC	800	−10	0.99	6.9	7.8	8

Table 9.6 Experimental results of the worst source current THDi_s of FS-HMPC under system parameter variation

		L_c			C_{PF}			L_{PF}		
		−10%	/	+10%	−10%	/	+10%	−10%	/	+10%
Worst THDi_{sx} (%)	Inductive loads	6.9	6.4	6.5	6.9	6.4	6.7	6.9	6.4	7.0
	Capacitive loads	8.2	7.8	8.1	7.9	7.8	8.3	7.9	7.8	8.0

(A) ***Multi-Quasi-Proportional-Resonant Control***: The compensation performance under inductive loads of the MQPRC with gain scheduling is shown in Figs. 9.6a, 9.7a and Table 9.5. The MQPRC takes about 70 ms to reach the steady-state after starting the TC*LC*-HAPF because of the slow dynamic response of the integration process. And the THDi_{sx} is about 6.5% at steady-state.

(B) ***Finite-Set Model Predictive Control***: The compensation performance under inductive loads of the FS-MPC with coupling capacitor voltage observer is shown in Figs. 9.6b, 9.7b and Table 9.5. The FS-MPC takes 8 ms to the steady-state after the TC*LC*-HAPF starts because of the directly optimal switching state selection. And the steady-state THDi_{sx} is about 7%, which is bigger than the MQPRC.

(C) ***Finite-Set Hybrid Model Predictive Control***: The compensation performance under inductive loads of the proposed FS-HMPC with coupling capacitor voltage observer is shown in Figs. 9.6c, 9.7c and Table 9.5. The proposal provides a fast transient response (T_{st} = 8 ms) when the TC*LC*-HAPF starts operation, which is the same as the FS-MPC because of the optimal switching state selection. And the steady-state THDi_{sx} is a bit better than the MQPRC, at about 6.4%.

In addition, the compensation performance under capacitive loads of proposed FS-HMPC with coupling capacitor voltage observer is shown in Figs. 9.6d, 9.7d and

Table 9.5. The settling time is the same as the inductive loads case ($T_{st} = 8$ ms) and the steady-state THDi_{sx} is reduced to about 7.5% from 22%, which proves the effectiveness of the proposal even under capacitive loads case.

Table also summarizes the experimental data of different control schemes under both inductive and capacitive loads, which concludes that the proposed FS-HMPC is the best control strategy for the TC*LC*-HAPF with the lowest steady-state error and fastest transient response.

Further, the experimental average switching frequency and efficiency comparison of FS-MPC and proposed FS-HMPC is given in Table 9.3, which is consistent with the simulated results. And the experimental results of the system parameters L_c, C_{PF} and L_{PF} vary +10% and −10% are given in Table 9.6. The results show that the proposed FS-HMPC is robust against model mismatches, which is also consistent with the simulation results.

9.7 Summary

A FS-HMPC for the TC*LC*-HAPF which is a combination of FS-MPC and hysteresis SMC is proposed in this chapter. It provides the lowest steady-state error and fastest transient response among the existing control strategy for the TC*LC*-HAPF. Also, the robustness is increased compared to the FS-MPC because of the merits of SMC. The stability and robustness analysis are presented bases on the discrete SMC theory, and the sliding coefficient design method is provided. Also, this chapter introduced a novel overall control scheme for the finite-set hybrid model predictive controlled TC*LC*-HAPF to operate under distorted and unbalanced grid voltage. Finally, the performance of the proposed control scheme is verified by simulation and experimental results.

References

1. C. Zheng, T. Dragičević, Z. Zhang, J. Rodriguez, F. Blaabjerg, Model predictive control of LC-filtered voltage source inverters with optimal switching sequence. IEEE Trans. Power Electron. **36**(3), 3422–3436 (2021)
2. H. Komurcugil, S. Biricik, S. Bayhan, Z. Zhang, Sliding mode control: Overview of its applications in power converters. IEEE Ind. Electron. Mag. **15**(1), 40–49 (2021)
3. C. Gong, W.-K. Sou, C.-S. Lam, Second-order sliding-mode current controller for LC-coupling hybrid active power filter. IEEE Trans. Ind. Electron. **68**(3), 1883–1894 (2021)
4. T. He, D.D.-C. Lu, L. Li, J. Zhang, L. Zheng, J. Zhu, Model-predictive sliding-mode control for three-phase AC/DC converters. IEEE Trans. Power Electron. **33**(10), 8982–8993 (2018)
5. T. He, M. Wu, D.D.-C. Lu, K. Song, J. Zhu, Model predictive sliding control for cascaded H-bridge multilevel converters with dynamic current reference tracking. IEEE Trans. Emerg. Sel. Topics Power Electron. **10**(2), 1409–1421 (2022)

6. S. Kwak, J.C. Park, Switching strategy based on model predictive control of VSI to obtain high efficiency and balanced loss distribution. IEEE Trans. Power Electron. **29**(9), 4554–4567 (2014)
7. C. Zheng, Z. Gong, X. Wu, T. Dragicevic, J. Rodriguez and F. Blaabjerg, Finite-set quasi-sliding mode predictive control of LC-filtered voltage source inverters. IEEE Trans. Ind. Electron. Early access (2022). https://doi.org/10.1109/TIE.2021.3135605
8. H. Özbay, *Introduction to Feedback Control Theory*, 1st ed. (CRC Press, Boca Raton, FL, 2019). https://doi.org/10.1201/9780203750117
9. A. Bartoszewicz, Discrete-time quasi-sliding-mode control strategies. IEEE Trans. Ind. Electron. **45**(4), 633–637 (1998)
10. K. Adamiak, Reference sliding variable based chattering-free quasi-sliding mode control. IEEE Access **8**, 133086–133094 (2020)
11. IEEE recommended practices and requirements for harmonic control in electrical power systems (2014). IEEE Standard 519-2014

Chapter 10
Second Order Sliding Model Observer Based Second Order Sliding Model Control for TC*LC*-HAPF

10.1 Introduction

In this chapter, a second-order sliding mode observer (SOSMO) based second-order sliding mode controller (SOSMC) for the TC*LC*-HAPF is proposed. Comparing to other observers, the SOSMO can be directly applied to nonlinear systems like TC*LC*-HAPF, and it is fast and accurate with good system robustness. The proposed SOSMO based SOSMC design ensures the TC*LC*-HAPF low steady-state error, fast dynamic response, fixed switching frequency, and good system robustness under system parameter variation and non-ideal grid conditions. First, the SOSMO for the TC*LC*-HAPF system model is proposed, and its parameter design is provided. Based on the SOSMO model, the SOSMC for the TC*LC*-HAPF is proposed and designed accordingly. Then, simulation studies and experiments on a three-phase TC*LC*-HAPF are carried out to verify the steady-state and dynamic performance of the proposed SOSMO based SOSMC design compared with the other existing current controllers: proportional-integral controller, hysteresis current controller and multi-quasi-proportional-resonant controller. Finally, the system robustness, unbalance compensation capability, and DC-link voltage control performance of the proposed SOSMO based SOSMC design under different non-ideal grid conditions are also verified by simulation and experimental results.

The contribution of this chapter is summarized as follows:

(1) Propose a SOSMO for real-time observation of the TC*LC* part's filtering capacitor voltage without additional sensor costs, which also simplifies the TC*LC*-HAPF system model for the SOSMC design.
(2) Propose a SOSMO based SOSMC for the TC*LC*-HAPF with low steady-state error, fast transient response, fixed switching frequency, and good system robustness under parameter variation and non-ideal grid conditions.
(3) Study parameter design of both the SOSMO and the SOSMC for the TC*LC*-HAPF to ensure its system stability and performance under system parameter variation and non-ideal grid conditions.

C.-S. Lam et al., *Advanced Controllers Design for Hybrid Active Power Filters*,
https://doi.org/10.1007/978-981-97-6847-9_10

(4) Verify the performance and robustness of the proposed SOSMO based SOSMC design for the TC*LC*-HAPF by simulation and experimental results in comparison with the PIC, HCC, and MQPRC.

10.2 Second-Order Sliding Mode Observer Design for TC*LC*-HAPF

From the circuit configuration of the three-phase TC*LC*-HAPF as shown in Fig. 1.5, the equivalent circuit system model of the three-phase TC*LC*-HAPF is shown in Fig. 10.1a, which can be derived as follows [1]:

$$L_c \frac{di_{cx}}{dt} = v_x - v_{cx} - v_{invx} \tag{10.1}$$

$$C_{PF} \frac{dv_{cx}}{dt} = i_{cx} - i_{gx} \tag{10.2}$$

$$L_{PF} \frac{di_{gx}}{dt} = G_x v_{cx} \tag{10.3}$$

Based on the system model of the three-phase TC*LC*-HAPF (10.1)–(10.3), the SOSMO is proposed in this part to obtain a real-time estimation of v_{cx} without using extra sensors, thus reducing the sensor cost. Moreover, it can simplify the SOSMC design, which will be discussed in Sect. 10.3.

The formulation of the SOSMO is designed as:

$$L_c \frac{d\hat{i}_{cx}}{dt} = v_x - \hat{v}_{cx} - v_{invx}{}^* + \lambda_{ox}\sqrt{|\hat{\sigma}_x|}\mathrm{sign}(\hat{\sigma}_x) \tag{10.4}$$

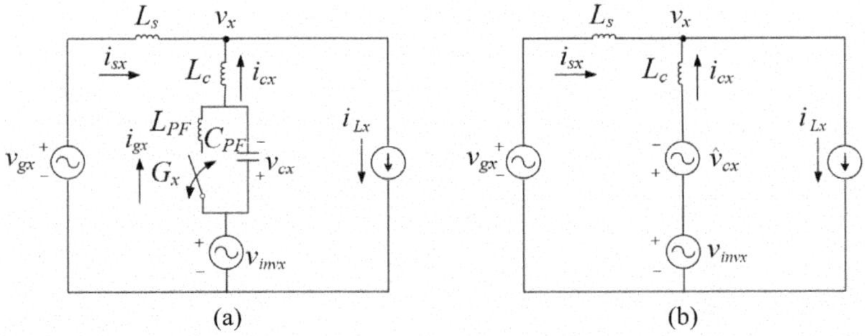

Fig. 10.1 Equivalent circuits of TC*LC*-HAPF with different models. **a** Original model. **b** SOSMO based simplified model

$$C_{PF}\frac{d\hat{v}_{cx}}{dt} = i_{cx} + \kappa_{ox}\text{sign}(\hat{\sigma}_x) \tag{10.5}$$

where $v_{invx}{}^{*}$ is the inverter voltage reference; $\hat{i}_{cx}$, $\hat{v}_{cx}$ are observed values of the compensating current and TC*LC* part's filtering capacitor voltage; λ_{ox} and κ_{ox} are the observer gains of the SOSMO; $\hat{}\sigma_x$ is the sliding variable of the SOSMO defined by:

$$\hat{\sigma}_x = i_{cx} - \hat{i}_{cx} \tag{10.6}$$

which is the observation error of the compensating current i_{cx}.

Subtracting (10.4) and (10.5) from (10.1) and (10.2), the error dynamics of the SOSMO is derived as:

$$\frac{d\hat{\sigma}_x}{dt} = -\frac{\lambda_{ox}}{L_c}\sqrt{|\hat{\sigma}_x|}\text{sign}(\hat{\sigma}_x)+\hat{\gamma}_x \tag{10.7}$$

$$\frac{d\hat{\gamma}_x}{dt} = -\frac{\kappa_{ox}}{L_c C_{PF}}\text{sign}(\hat{\sigma}_x) - \frac{i_{gx}}{L_c C_{PF}} \tag{10.8}$$

where $\hat{\gamma}_x$ is the auxiliary variable of the TC*LC* part's filtering capacitor voltage v_c observation error, which is defined by:

$$\hat{\gamma}_x = \frac{\hat{v}_{cx} - v_{cx}}{L_c} \tag{10.9}$$

Defining M_{ox} is the upper bound of $|i_{gx}|$, which can be obtained when the G_x is always on state. In the case $G_x = 1$, the *s*-domain expression of i_{gx} can be derived from (10.1)–(10.3), which is given as:

$$i_{gx}(s) = \frac{v_x(s) - v_{invx}(s)}{s^3 L_c L_{PF} C_{PF} + (L_c + L_{PF})s} \tag{10.10}$$

Based on the principle of Laplace transform, the magnitude of i_{gx} at each harmonic frequency is derived as:

$$\left|i_{gx}(jn\omega_1)\right| = \left|\frac{v_x(jn\omega_1) - v_{invx}(jn\omega_1)}{(jn\omega_1)^3 L_c L_{PF} C_{PF} + jn\omega_1(L_c + L_{PF})}\right| \tag{10.11}$$

where ω_1 is the fundamental grid frequency. From (10.11), an upper bound of the maximum magnitude of i_{gx} can be found:

$$
\begin{aligned}
\left|i_{gx}\right| &\le \sum_{n}^{h}\left|i_{gx}(jn\omega_1)\right| \\
&\le \sum_{n}^{h}\frac{V_{xn}+v_{dc}}{\left|(jn\omega_1)^3 L_c L_{PF} C_{PF} + jn\omega_1 L_{PF} + jn\omega_1 L_c\right|} \\
&\le \sum_{n}^{h}\frac{\overline{V}_{xn}+\overline{v}_{dc}}{\left|(jn\underline{\omega}_1)^3 \underline{L}_c \underline{L}_{PF} \underline{C}_{PF} + jn\underline{\omega}_1 \underline{L}_{PF} + jn\underline{\omega}_1 \underline{L}_c\right|} \\
&= M_{ox}
\end{aligned}
\tag{10.12}
$$

where h is the maximum harmonic order under consideration, V_{xn} is the magnitude of the n-th order load voltage, the upper and under lines "‾" and "_" denotes the maximum and minimum values of the corresponding parameters, which can be easily obtained from the data sheets of the inductors and capacitors, and IEEE Standards [2, 3].

To guarantee the performance of SOSMO under system parameter variation, distorted grid voltage, and grid frequency variation, the worst-case M_{ox} under these conditions can be expressed as:

$$
M_{ox} = \sum_{n}^{h}\frac{\overline{V}_{xn}+\overline{v}_{dc}}{\left|(jn\underline{\omega}_1)^3 \underline{L}_c \underline{L}_{PF} \underline{C}_{PF} + jn\underline{\omega}_1 \underline{L}_{PF} + jn\underline{\omega}_1 \underline{L}_c\right|} \tag{10.13}
$$

Therefore, the filtering capacitor voltage observation error $\hat{}\gamma_x$ converges to zero under system parameter variation, distorted grid voltage, and grid frequency variation, if the SOSMO observer gain parameters κ_{ox} and λ_{ox} satisfy the following stability condition [4]:

$$
\kappa_{ox} > M_{ox} \tag{10.14}
$$

$$
\lambda_{ox} > \overline{L}_c\sqrt{\kappa_{ox} + \frac{M_{ox}}{\underline{L}_c \underline{C}_{PF}}} \tag{10.15}
$$

10.3 Observer Based Second-Order Sliding Mode Controller Design for TC*LC*-HAPF

Based on the SOSMO, the system model Eqs. (10.1)–(10.3) can be rewritten as:

$$
L_c\frac{di_{cx}}{dt} = v_x - \hat{v}_{cx} - v_{invx} \tag{10.16}
$$

where the equivalent circuit of the TCLC-HAPF with the SOSMO based simplified model is shown in Fig. 10.1b.

Based on the simplified model (10.16), the proposed SOSMC design in this chapter is given as:

$$v_{invx}{}^{*} = \lambda_x \sqrt{|\sigma_x|}\,\mathrm{sign}(\sigma_x) + \kappa_x \int \mathrm{sign}(\sigma_x) - \hat{v}_{cx} \tag{10.17}$$

where λ_x and κ_x are the control parameters of the SOSMC. σ_x is the sliding variable which is defined as:

$$\sigma_x = i_{cx}{}^{*} - i_{cx} \tag{10.18}$$

where $i_{cx}{}^{*}$ is the reference compensating current.

Replace the v_{invx} by its reference value $v_{invx}{}^{*}$ (substituting (10.17) into (10.16)), the dynamic of the sliding variable (compensating current error, σ_x) can be derived by taking a time derivative of (10.18), which is written as:

$$\frac{d\sigma_x}{dt} = -\frac{\lambda_x}{L_c}\sqrt{|\sigma_x|}\,\mathrm{sign}(\sigma_x) + \gamma_x \tag{10.19}$$

$$\frac{d\gamma_x}{dt} = -\frac{\kappa_x}{L_c}\mathrm{sign}(\sigma_x) + \delta_x \tag{10.20}$$

where γ_x is an auxiliary variable. δ_x stands for the total 'perturbation' of the system (10.19), (10.20):

$$\delta_x = \frac{d^2 i_{cx}{}^{*}}{dt^2} - \frac{1}{L_c}\frac{dv_x}{dt} \tag{10.21}$$

Defining M_x as the constant upper boundary of $|\delta_x|$, which can be calculated by:

$$\begin{aligned}
|\delta_x| &= \left|\frac{d^2 i_{cx}{}^{*}}{dt^2} - \frac{1}{L_c}\frac{dv_x}{dt}\right| \\
&\leq \left|\sum_{n=1}^{h} I_{cxn}(n\omega_1)^2 - \sum_{n=1}^{h}\frac{1}{L_c}V_{xn}(n\omega_1)\right| \\
&\leq \left|\sum_{n=1}^{h} \overline{I}_{Lxn}(n\overline{\omega}_1)^2 - \sum_{n=1}^{h}\frac{1}{\underline{L}_n}\underline{V}_{xn}(n\underline{\omega}_1)\right| \\
&= M_x
\end{aligned} \tag{10.22}$$

where I_{cxn} is the magnitude of the n-th harmonic compensating current, which is always smaller than the maximum available value of n-th harmonic load current I_{Lxn}.

Table 10.1 Parameter design summary of the proposed SOSMO based SOSMC design

	SOSMO	SOSMC
α_{ox} or κ_x	$\kappa_{ox} > M_{ox}$	$\kappa_x > \overline{L}_c M_x$
λ_{ox} or λ_x	$\lambda_{ox} > \overline{L}_c \sqrt{\kappa_{ox} + \frac{M_{ox}}{\underline{L}_c \underline{C}_{PF}}}$	$\lambda_x > \overline{L}_c \sqrt{\kappa_x + M_x}$
M_{ox} or M_x	$M_{ox} = \sum_{n}^{h} \frac{\overline{V}_{xn} + \overline{v}_{dc}}{\lvert (jn\underline{\omega}_1)^3 \underline{L}_c \underline{L}_{PF} \underline{C}_{PF} + jn\underline{\omega}_1 \underline{L}_{PF} + jn\underline{\omega}_1 \underline{L}_c \rvert}$	$M_x = \left\lvert \sum_{n=1}^{h} \overline{I}_{Lxn}(n\overline{\omega}_1)^2 - \sum_{n=1}^{h} \frac{1}{\underline{L}_c} \underline{V}_{xn}(n\underline{\omega}_1) \right\rvert$

Similar as the SOSMO design procedure, to guarantee the performance of the SOSMC under load changes, system parameter variation, distorted grid voltage, and grid frequency variation, the worst-case calculation of M_x under these conditions can be expressed as:

$$M_x = \left| \sum_{n=1}^{h} \overline{I}_{Lxn}(n\overline{\omega}_1)^2 - \sum_{n=1}^{h} \frac{1}{\underline{L}_c} \underline{V}_{xn}(n\underline{\omega}_1) \right| \tag{10.23}$$

Therefore, the compensating current error will converge to zero under load changes, system parameter variation, distorted grid voltage, and grid frequency variation, if its designed control gain parameters satisfy [4]:

$$\kappa_x > \overline{L}_c M_x \tag{10.24}$$

$$\lambda_x > \overline{L}_c \sqrt{\kappa_x + M_x} \tag{10.25}$$

In summary, the design guidelines of the proposed SOSMO based SOSMC design for the TC*LC*-HAPF are shown in Table 10.1.

10.4 Digital Implementation of SOSMO Based SOSMC

Based on (10.7) (10.8) and (10.17), the control block diagram of the SOSMO based SOSMC is proposed in Fig. 10.2. And the digital implementation of the proposed SOSMO based SOSMC design for the TC*LC*-HAPF is shown in Fig. 10.3. The blue area in Fig. 10.3 is defined as second-order sliding mode blocks (SOSMB), which contains an absolute value function (if/else operation), a square root operation, a multiplication, and two sign functions (also if/else operation). The grey area is the digital integrator (DI) based on trapezoidal rule [5, 6], which has two addition/subtraction operations, and a multiplication operation. Besides of the SOSMA and DI, the other operation stages for the SOSMC part and SOSMO part are simply addition, subtraction, and multiplication operations, which can be solved by one cycle of the DSP with hardware multiplier [7].

Table 10.2 summarizes the required DSP's digital operations of the proposed SOSMO based SOSMC design, which shows the implementation complexity of the proposed controller. The only multiple-step operation is the square root calculation,

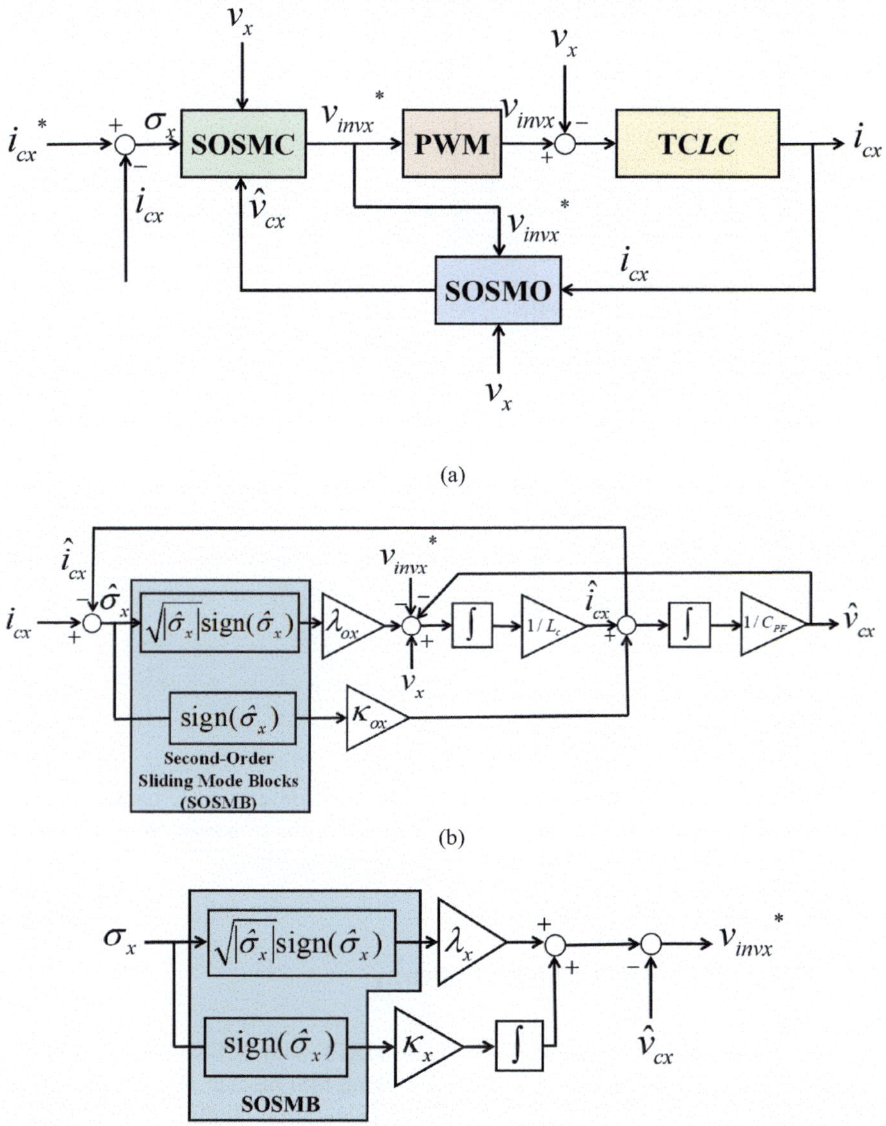

Fig. 10.2 Control block diagram of SOSMO based SOSMC. **a** Global view. **b** SOSMO. **c** SOSMC

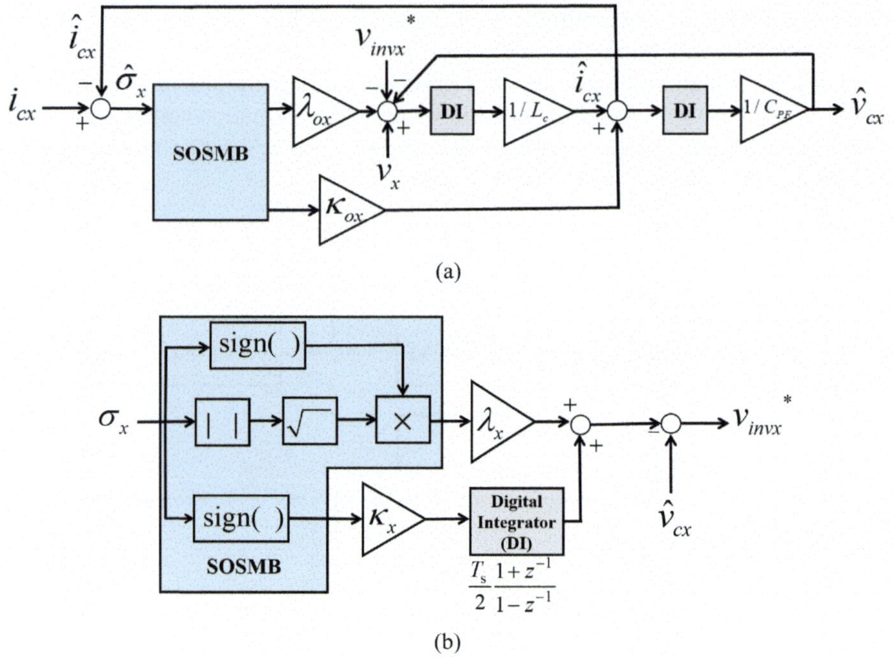

Fig. 10.3 Diagrams of the proposed SOSMO based SOSMC design digital implementation for the TC*LC*-HAPF: **a** SOSMO part; **b** SOSMC part

Table 10.2 DSP operations requirement of the proposed SOSMO based SOSMC design

Operation	±	*	If/else	√
SOSMC	4	4	3	1
SOSMO	8	7	3	1

which can be efficiently solved by applying the well-developed C packages such as IQmath of TI DSP [7]. Therefore, from Table 10.2, the proposed SOSMO based SOSMC design can be easily applied in real-time applications.

10.5 Overall Control Block Diagram

The overall control block diagram of the TC*LC*-HAPF with SOSMO based SOSMC is shown in Fig. 10.4 which consists of the following four blocks: TC*LC* part control block, active inverter part instantaneous power calculation block, DC-link voltage control block, and final compensating current with SOSMO based SOSMC block.

For the TC*LC* part control block, as shown in Fig. 10.4 the instantaneous loading reactive power is calculated by using the single-phase instantaneous *p*–*q* theory [8],

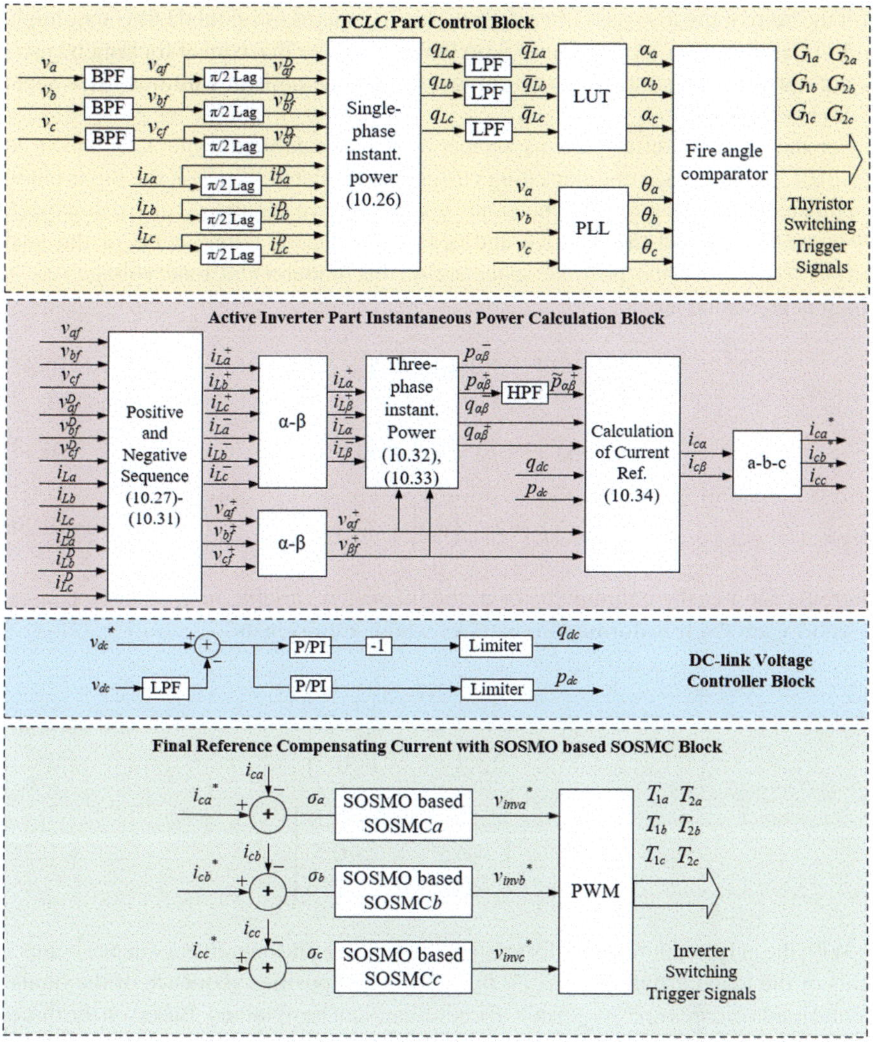

Fig. 10.4 Overall control block diagram of the SOSMO based SOSMC TCLC-HAPF

in which the instantaneous loading reactive power q_{Lx} is defined by:

$$\begin{bmatrix} q_{La} \\ q_{Lb} \\ q_{Lc} \end{bmatrix} = \begin{bmatrix} v_{bf} \cdot i_{Lc}^{D} - v_{cf}^{D} \cdot i_{Lb} \\ v_{cf} \cdot i_{La}^{D} - v_{af}^{D} \cdot i_{Lc} \\ v_{af} \cdot i_{Lb}^{D} - v_{bf}^{D} \cdot i_{La} \end{bmatrix} \tag{10.26}$$

where v_{xf} is the fundamental load voltage extracted by bandpass filter (BPF). $\overline{q}_{Lx}$ is the dc component of the instantaneous loading reactive power q_{Lx}, which are obtained with the help of the low-pass filter (LPF). The $\overline{q}_{Lx}$ is used to calculate the firing angle (α_x) of the thyristors based on a lookup table. Then the thyristor switching triggers G_x are obtained from the fire angle comparator, in comparison with the phase angle of the load voltage θ_x.

For the active inverter part instantaneous power calculation block as shown in Fig. 10.4, the reference compensating current is calculated based on the three-phase instantaneous p–q theory [8]. The positive and negative sequence calculation block [9] is used to extract the positive and negative sequence components of the load current ($i_{Lx}{}^{+}$, $i_{Lx}{}^{-}$) and positive sequence of the fundamental load voltage ($v_{xf}{}^{+}$), which is presented as:

$$\left[i_{Lx}{}^{+}\right] = T_2 \cdot [i_{Lx}] - T_1 \cdot \left[i_{Lx}^{D}\right] \tag{10.27}$$

$$\left[i_{Lx}{}^{-}\right] = T_2 \cdot [i_{Lx}] + T_1 \cdot \left[i_{Lx}^{D}\right] \tag{10.28}$$

$$\left[v_{xf}{}^{+}\right] = T_2 \cdot \left[v_{xf}\right] - T_1 \cdot \left[v_{xf}^{D}\right] \tag{10.29}$$

where $[\cdot]$ denotes the column vector of the bracketed variable in a, b, and c phases. T_1 and T_2 are the transformation matrices which are given as:

$$T_1 = \frac{1}{2\sqrt{3}} \begin{bmatrix} 0 & 1 & -1 \\ -1 & 0 & -1 \\ 1 & -1 & 0 \end{bmatrix} \tag{10.30}$$

$$T_2 = \frac{1}{3} \begin{bmatrix} 1 & -0.5 & -0.5 \\ -0.5 & 1 & -0.5 \\ -0.5 & -0.5 & 1 \end{bmatrix} \tag{10.31}$$

With the help of Clarke transformation, the positive and negative sequence components of the load current ($i_{L\alpha}{}^{+}$, $i_{L\beta}{}^{+}$, $i_{L\alpha}{}^{-}$, $i_{L\beta}{}^{-}$) and positive sequence of the fundamental load voltage ($v_{\alpha f}{}^{+}$, $v_{\beta f}{}^{+}$) in α-β coordinate can be obtained. Based on the three-phase instantaneous p–q theory, the three-phase instantaneous positive sequence and negative sequence loading active powers and reactive powers ($p_{\alpha\beta}{}^{+}$, $p_{\alpha\beta}{}^{-}$, $q_{\alpha\beta}{}^{+}$, $q_{\alpha\beta}{}^{-}$) can be calculated by:

$$\begin{bmatrix} p_{\alpha\beta}{}^{+} \\ q_{\alpha\beta}{}^{+} \end{bmatrix} = \begin{bmatrix} v_{\alpha f}{}^{+} & v_{\beta f}{}^{+} \\ -v_{\beta f}{}^{+} & v_{\alpha f}{}^{+} \end{bmatrix} \cdot \begin{bmatrix} i_{L\alpha}{}^{+} \\ i_{L\beta}{}^{+} \end{bmatrix} \tag{10.32}$$

$$\begin{bmatrix} p_{\alpha\beta}{}^{-} \\ q_{\alpha\beta}{}^{-} \end{bmatrix} = \begin{bmatrix} v_{\alpha f}{}^{+} & v_{\beta f}{}^{+} \\ -v_{\beta f}{}^{+} & v_{\alpha f}{}^{+} \end{bmatrix} \cdot \begin{bmatrix} i_{L\alpha}{}^{-} \\ i_{L\beta}{}^{-} \end{bmatrix} \tag{10.33}$$

Then the three-phase instantaneous positive sequence loading active power harmonics component ($\tilde{p}_{\alpha\beta}{}^{+}$) can be obtained by passing through $p_{\alpha\beta}{}^{+}$ to the high pass filter (HPF). Then the $\tilde{p}_{\alpha\beta}{}^{+}$, $p_{\alpha\beta}{}^{-}$, $q_{\alpha\beta}{}^{+}$, $q_{\alpha\beta}{}^{-}$ (containing the unbalance, harmonic, and reactive power components of the loading) and the DC-link voltage control signals p_{dc} and q_{dc} are sent to the reference compensating current calculation block to calculate the reference compensating currents $(i_{c\alpha}, i_{c\beta})$ in α-β coordinate, which is given as:

$$\begin{bmatrix} i_{c\alpha}{}^{*} \\ i_{c\beta}{}^{*} \end{bmatrix} = \frac{1}{v_{\alpha f}^{2}{}^{+} + v_{\beta f}^{2}{}^{+}} \begin{bmatrix} v_{\alpha f}{}^{+} & -v_{\beta f}{}^{+} \\ v_{\beta f}{}^{+} & v_{\alpha f}{}^{+} \end{bmatrix} \cdot \begin{bmatrix} \tilde{p}_{\alpha\beta}{}^{+} + p_{\alpha\beta}{}^{-} + p_{dc} \\ q_{\alpha\beta}{}^{+} + q_{\alpha\beta}{}^{-} - q_{dc} \end{bmatrix} \tag{10.34}$$

Then the reference compensating currents $i_{cx}{}^{*}$ in a-b-c coordinate can be obtained by using the inverse matrix of Clarke transformation in α-β coordinates.

The DC-link voltage control block is used to stabilize the DC-link voltage. The error signal of DC-link voltage is sent to two proportional-integral (PI) controllers to generate the corresponding DC-link voltage control signals p_{dc} and q_{dc}. The p_{dc} and q_{dc} will be sent to the reference compensating current calculation block for calculating the reference compensating currents in α-β coordinate.

For the final reference compensating current with SOSMO based SOSMC block, the compensating current error σ_x signal of each phase goes through the proposed SOSMO based SOSMC design and generates the reference inverter voltage $v_{invx}{}^{*}$. Then, the switching trigger signals (T_{1x}, T_{2x}) of the inverter are obtained from the PWM block.

10.6 Simulation Results

For verifying the validity and effectiveness of the proposed SOSMO based SOSMC design for the T*CLC*-HAPF, MATLAB/Simulink based simulation studies were carried out. The system, T*CLC*-HAPF, loads and different controllers parameters are presented in Table 10.3, where f_g and f_{sw} are the system grid and active inverter switching frequencies, K_p and K_i are proportional and integral gains of PIC [1], H is the hysteresis band of HCC [11–15], K_p, K_{r1}, K_{r5}, K_{r7}, ω_c are proportional gain, fundamental, 5th and 7th harmonics order resonant gains, and bandwidth of resonant frequency of the MQPRC [17]. The parameter design methodology of the PIC, HCC and MQPRC can be referred to [1, 11–15, 17].

First, to verify the effectiveness of the proposed SOSMO, simulation results of the actual T*CLC* part's filtering capacitor voltage v_{cx} and its observation $\hat{v}_{cx}$ values are compared under the T*CLC*-HAPF compensating different loads. The simulation results under inductive and capacitive nonlinear load conditions are presented in Fig. 10.5, the maximum steady-state observation errors are 8.5 V (inductive nonlinear load condition) and 9.9 V (capacitive nonlinear load condition), which are less than 1% under both load conditions.

Moreover, the simulation results of the SOSMC for the T*CLC*-HAPF with directly measured, observed, and without capacitor voltages v_{cx} under different T*CLC* part

Table 10.3 Parameters of system, TC*LC*-HAPF, loads and different controllers

	Parameters	Value
System grid	v_{gx}, f_g	Simulation 11 kV, 50 Hz Experiment 110 V, 50 Hz
Active inverter part	v_{dc}, f_{sw}	Simulation 5 kV, 10 kHz Experiment 60 V, 10 kHz
TC*LC* part	L_c, L_{PF}, C_{PF}	5 mH, 30 mH, 160 μF
Sampling time	T_s	100 μs
Non-linear load	R_{NL}, L_{NL}	50 Ω, 30 mH
Inductive linear load	R_{LL}, L_{LL}	15 Ω, 70 mH
Capacitive linear load	R_{LL}, C_{LL}	25 Ω, 140 μF
PIC	K_p, K_i	145, 25,970
HCC	H	0.125
MQPRC	$K_p, K_{r1}, K_{r5}, K_{r7}, \omega_c$	17.6, 850, 1220, 950, 6.28
SOSMO based SOSMC	$\lambda_{ox}, \kappa_{ox}, \lambda_x, \kappa_x$	10, 3000, 200, 35,000

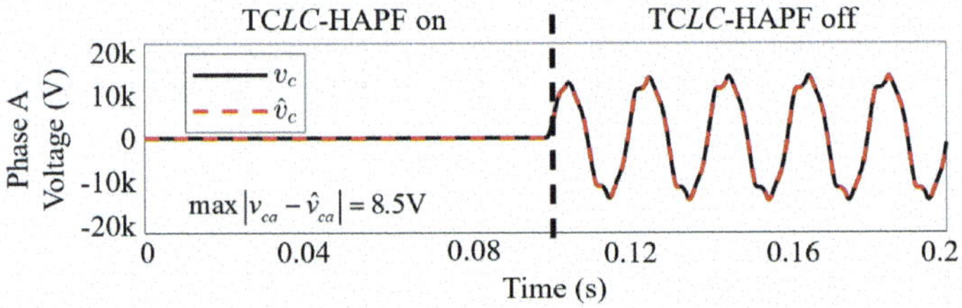

(a)

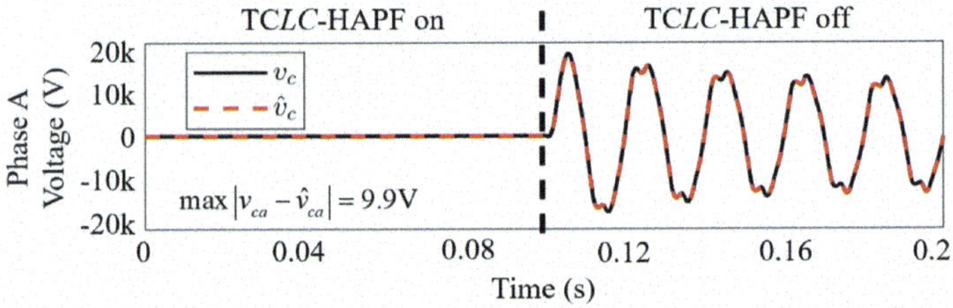

(b)

Fig. 10.5 Simulation results of the TC*LC* part's filtering capacitor voltage v_{cx} and its SOSMO observation under different load conditions. **a** Inductive nonlinear load condition. **b** Capacitive nonlinear load condition

firing angles (different load conditions) are given in Fig. 10.6. The SOSMC with SOSMO observed capacitor voltage obtains approximately the same simulated steady-state and dynamic performances as the SOSMC with directly sensing the v_{cx} signals, and better performances than the SOSMC without v_{cx}. Moreover, from Figs. 10.5 and 10.6, the observation error of v_{cx} of the proposed SOSMO is small, and

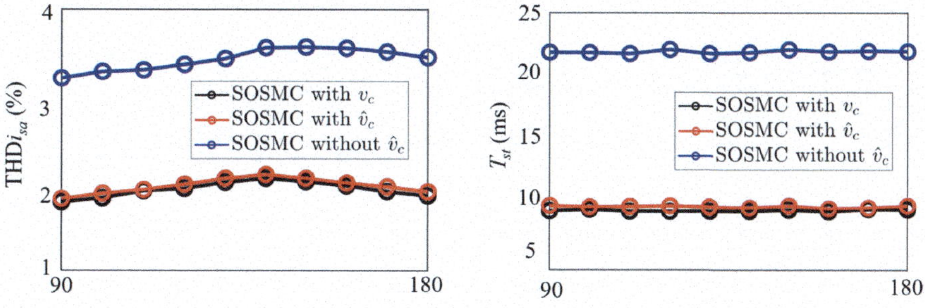

Fig. 10.6 Simulation results of the SOSMC controlled TC*LC*-HAPF system with directly measured, SOSMO observed, and without v_{cx} under different TC*LC* part firing angles. **a** System current THDi_{sx} with TC*LC* part firing angle changes (90° → 180°). **b** Settling time T_{st} with TC*LC* part firing angle changes (90° → 180°)

the SOSMO based SOSMC for the TC*LC*-HAPF obtains approximately the same simulation results as the SOSMC case with directly sensing the v_{cx} signals with three additional sensors, which clearly shows that we can simply use $\hat{v}_{cx}$ to substitute v_{cx}, thus saving the sensor cost.

To verify the robustness and disturbance rejection capability of the proposed SOSMO based SOSMC design, simulation studies under $\pm 15\%$ system parameters variation (L_c, C_{PF}, and L_{PF} varying simultaneously), ± 1.5 Hz grid frequency variation, and maximum 15% grid voltage distortion of the TC*LC*-HAPF are also carried out and the simulation results are shown in Fig. 10.7, in comparison with PIC, HCC, and MQPRC.

From Fig. 10.7, the proposed SOSMO based SOSMC design shows superior system robustness and compensation performances under system parameters variation, grid frequency variation and grid voltage distortion, comparing to the PIC, HCC, and MQPRC.

10.7 Experimental Results

Due to the limitation of the laboratory facilities, a scale down small rating 110V–5kVA three-phase three-wire TC*LC*-HAPF experimental prototype was built in the laboratory, which has been described in Chap. 6. The system, TC*LC*-HAPF, loads and different controllers parameters are presented in Table 10.3. To demonstrate the superior control performance of the proposed SOSMO based SOSMC design for the TC*LC*-HAPF, a comparative study of the compensation performances of the TC*LC*-HAPF with different current controllers will be presented. The PIC [1], HCC [11–15] and MQPRC [17] are applied to the TC*LC*-HAPF for perform a comparison with the proposed SOSMO based SOSMC design.

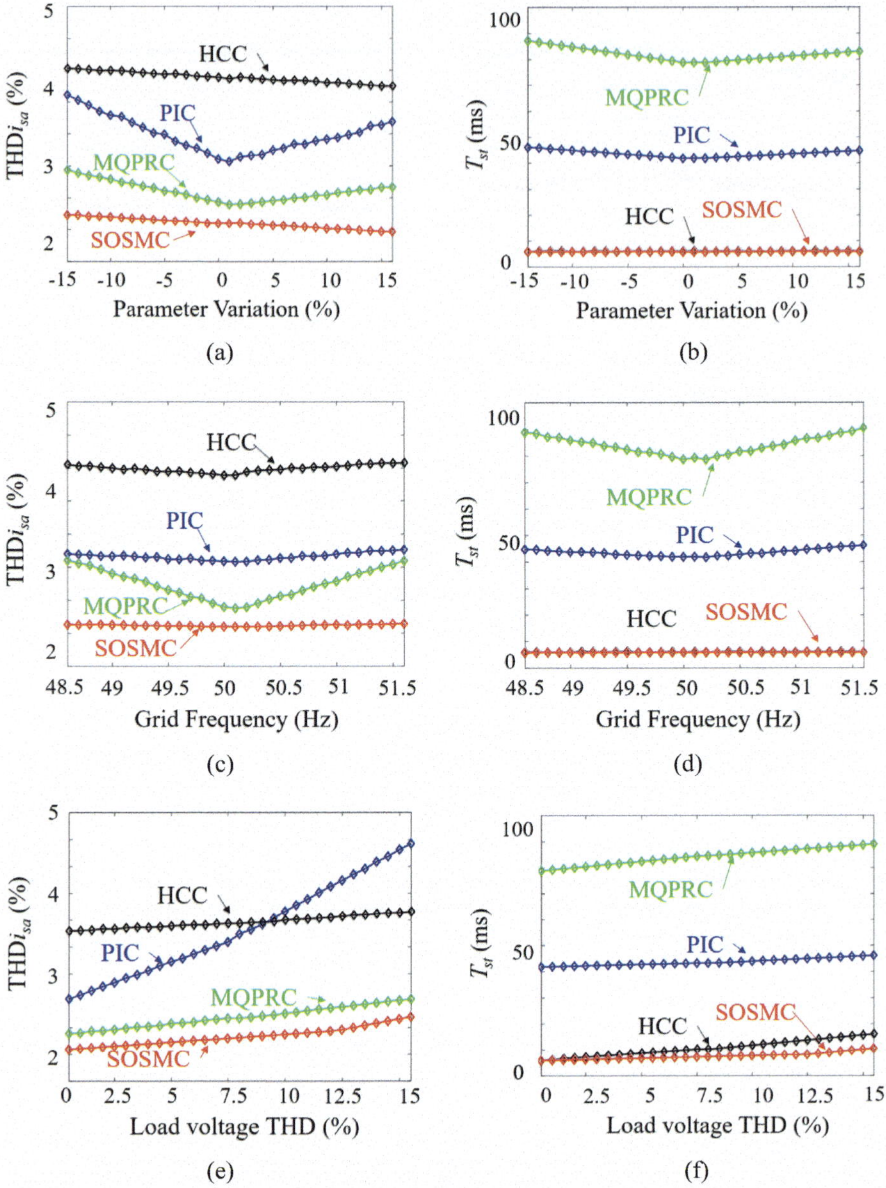

Fig. 10.7 Compensation results among PIC, HCC, MQPRC and the proposed SOSMO based SOSMC design for the TC*LC*-HAPF under system parameter (L_c, C_{PF}, and L_{PF}) variation and different grid conditions: **a** THDi_{sa} under system parameter variation; **b** T_{st} under system parameter variation; **c** THDi_{sa} under grid frequency variation; **d** T_{st} under grid frequency variation; **e** THDi_{sa} under different grid voltage distortion; **f** T_{st} under different grid voltage distortion

The experimental results of the TC*LC*-HAPF under different load conditions with different controllers: PIC, HCC, MQPRC, and the proposed SOSMO based SOSMC design are presented in Figs. 10.8, 10.9, 10.10 and 10.11. High order harmonic spectrum of the compensated system current with the HCC and the proposed SOSMO based SOSMC design are demonstrated in Fig. 10.12. The steady-state and dynamic performance indexes of the experimental results with different controllers are summarized in Table 10.4.

(A) ***Proportional-Integral Controller***: Figs. 10.8a, 10.9a and 10.11a show the compensating performance of the PIC controlled TC*LC*-HAPF. With the capacitor voltage v_{cx} sensors based feedforward algorithm [1], the PIC provides a satisfactory steady-state compensation performance with the worst compensated THDi_{sx} of 7.8%. However, the integrator characteristic of the linear integrator parts leads to a slow transient response up to about T_{st} = 50 ms. Besides, the sensors' cost of the PIC is the highest among those controllers.

(B) ***Hysteresis Current Controller***: Figs. 10.8b, 10.9b and 10.11b show the compensating performance of the TC*LC*-HAPF with HCC. According to the analysis results

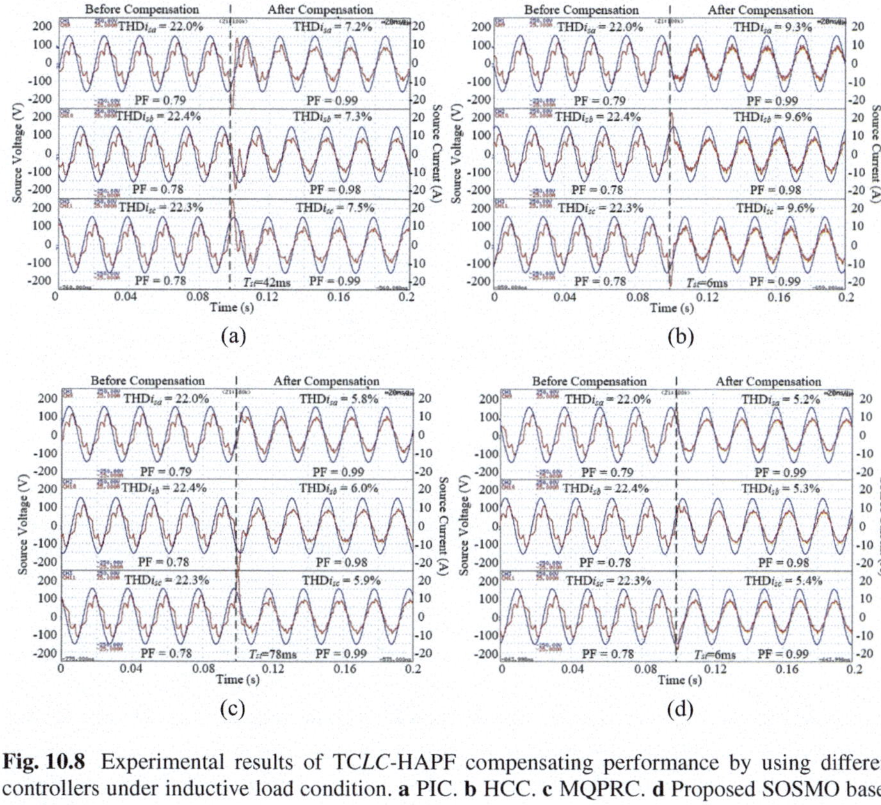

Fig. 10.8 Experimental results of TC*LC*-HAPF compensating performance by using different controllers under inductive load condition. **a** PIC. **b** HCC. **c** MQPRC. **d** Proposed SOSMO based SOSMC

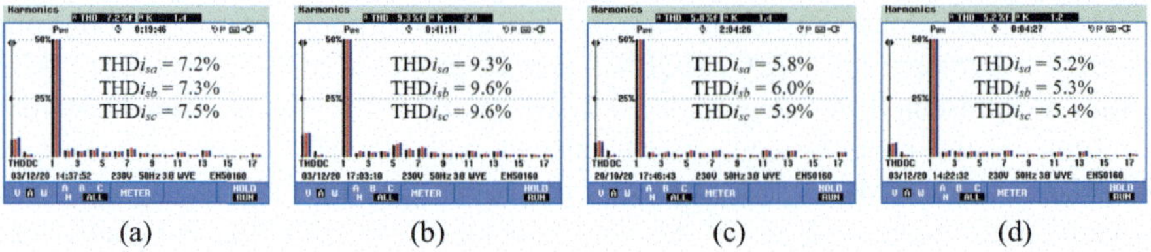

Fig. 10.9 Experimental results of harmonic spectrum of system current after TC*LC*-HAPF compensation under inductive load condition with different current controllers. **a** PIC. **b** HCC. **c** MQPRC. **d** Proposed SOSMO based SOSMC

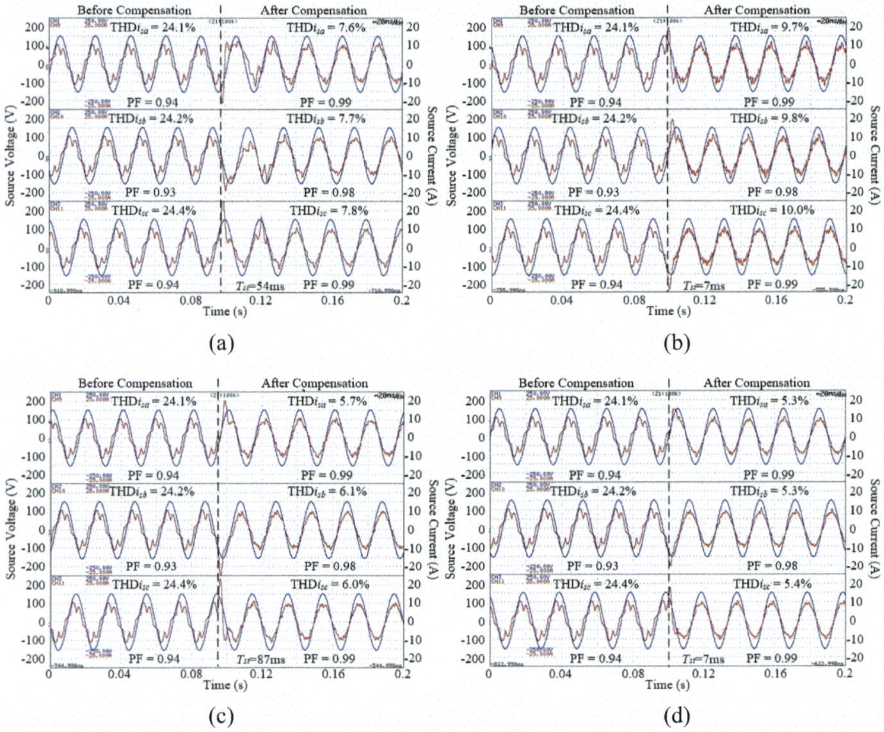

Fig. 10.10 Experimental results of TC*LC*-HAPF compensating performance by using different controllers under capacitive load condition. **a** PIC. **b** HCC. **c** MQPRC. **d** Proposed SOSMO based SOSMC

in [10], the HCC is equivalent to a proportional controller with an ideally infinite proportional gain. Thus, it provides a large control bandwidth, which yields fast transient response (T_{st} = 6–7 ms), as shown in Figs. 10.8b and 10.11b. However, the hysteresis band modulation algorithm leads to a varying switching frequency, which generates undesired high-order harmonics. The worst compensated THDi_{sx} is about 10%, which is the highest among all the controllers.

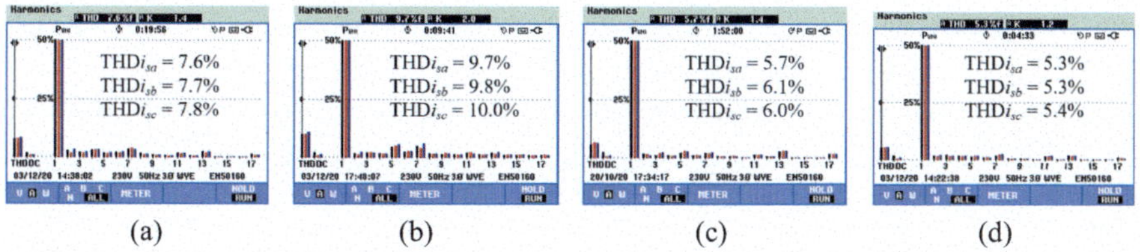

Fig. 10.11 Experimental results of harmonic spectrum of system current after TC*LC*-HAPF compensation under capacitive load condition with different current controllers. **a** PIC. **b** HCC. **c** MQPRC. **d** Proposed SOSMO based SOSMC

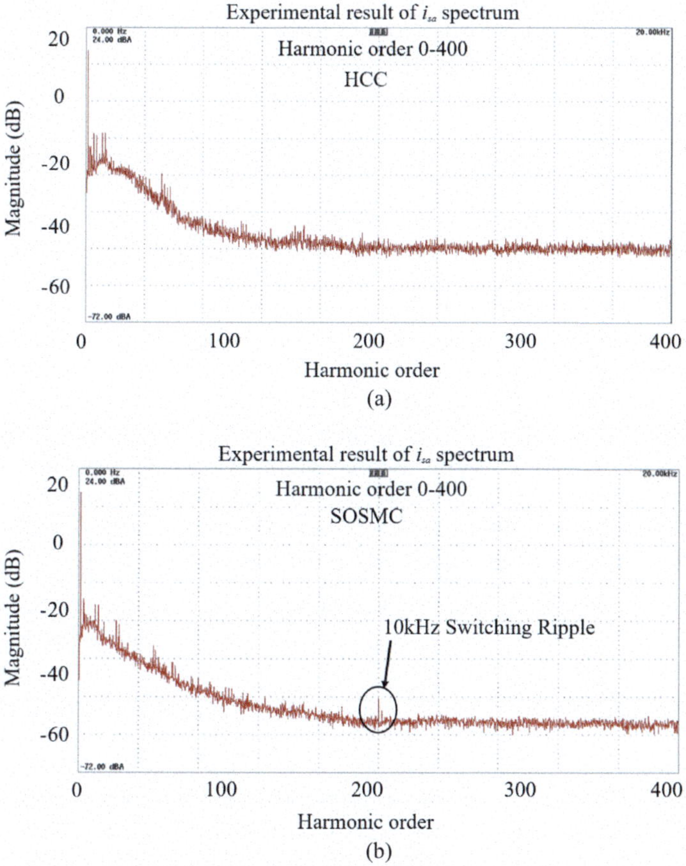

Fig. 10.12 High-order harmonic spectrum of the system current compensated by the TC*LC*-HAPF with different controllers. **a** HCC. **b** SOSMO based SOSMC

Table 10.4 Experimental results before and after TC*LC*-HAPF compensation by using different controllers

		P_{sx} (W)	Q_{sx} (var)	PF	T_{st} (ms)	Worst THDi_{sx} (%)
Inductive Loads	Before Compensation	610	450	0.78	/	22.4
	PIC	660	10	0.99	42	7.5
	HCC	670	40	0.98	6	9.6
	MQPRC	660	10	0.99	78	6.0
	SOSMO based SOSMC	660	10	0.99	6	5.4
Capacitive Loads	Before Compensation	740	−220	0.93	/	24.4
	PIC	770	−10	0.99	54	7.8
	HCC	770	−20	0.99	7	10.0
	MQPRC	770	−10	0.99	87	6.1
	SOSMO based SOSMC	770	−10	0.99	7	5.4

(C) ***Multi-Quasi-Proportional Resonant Controller***: Figs. 10.8c, 10.9c and 10.11c show the compensating performance of the TC*LC*-HAPF with MQPRC. As the MQPRC provides resonant peaks at the considered harmonic frequency orders, thus the steady-state error of the selected harmonics can be reduced to a very low level (worst THDi_{sx} = 6.1%). However, the second-order resonant parts also lead to a slow transient response. As shown in Figs. 10.8c and 10.11c, the transient response speed of MQPRC is the slowest (T_{st} = 78 ms for inductive load, T_{st} = 87 ms for capacitive load) among all the controllers in the comparative study.

(D) ***Second-Order Sliding Mode Observer based Second-Order Sliding Mode Controller***: Figs. 10.8d, 10.9d and 10.11d show the compensating performance of the TC*LC*-HAPF with the SOSMO based SOSMC. From the high-order sliding mode control theory [18–20], the SOSMO based SOSMC maintains the fast response speed (T_{st} = 6 ms for inductive load, T_{st} = 7 ms for capacitive load) with the best steady-state compensating performance (THDi_{sx} = 5.4% for both load conditions), as shown in Figs. 10.8d and 10.11d. Moreover, as the proposed SOSMO based SOSMC design generates a continuous controller output, a fixed switching frequency can be achieved.

The high order harmonic spectrum of the system current with the HCC and the proposed SOSMO based SOSMC are demonstrated in Fig. 10.12. From Fig. 10.12a, the varying switching frequency of HCC leads to a wide range of switching ripple, which is about 10 dB higher harmonic magnitude of the system current at high frequency range. On the contrary, the switching ripple of the SOSMO based SOSMC is mainly emerged at the fixed switching frequency 10 kHz, and thus easily reduced to a very low level (maximum −48 dB) by the well-designed output inductor, as shown in Fig. 10.12b.

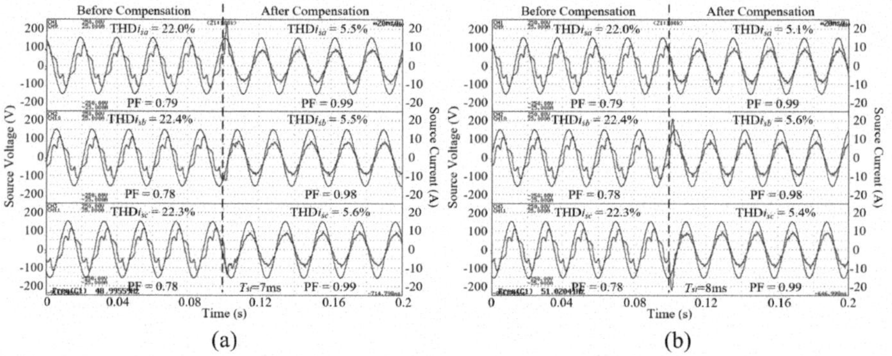

Fig. 10.13 Experimental results of TC*LC*-HAPF with proposed SOSMO based SOSMC under grid frequency variation conditions. **a** 49 Hz. **b** 51 Hz

To verify the effectiveness and robustness of the proposed controller under frequency variation, the experimental results of the TC*LC*-HAPF with SOSMO based SOSMC under frequency variation are provided. According to the IEEE Standard 446-1995 [2], the maximum frequency variation tolerance of the 50 Hz distribution power system is +/−0.5 Hz. In this chapter, a more severe frequency variation of +/−1 Hz is taken into consideration. The corresponding experimental results are shown in Fig. 10.13, the compensation performance is approximately the same under 49 and 51 Hz grid frequency, which verifies the robustness of the proposed SOSMO based SOSMC design under grid frequency variation.

To verify the effectiveness of the proposed TC*LC*-HAPF with SOSMO based SOSMC under distorted grid voltage condition, the experimental results of the TC*LC*-HAPF with the proposed SOSMO based SOSMC design under distorted grid voltage condition THDv_x = 9.3%, which does not satisfy the IEEE STD 519-2014 [3] of voltage THD < 8%, are demonstrated in Fig. 10.14. The system current THDi_{sx} can be compensated from 24.6 to 6.2% with a settling time of T_{st} = 8 ms, which verifies that the TC*LC*-HAPF with SOSMO based SOSMC can perform satisfactory compensation even under distorted grid voltage condition.

The experimental result of the TC*LC*-HAPF with the proposed SOSMO based SOSMC design under unbalanced nonlinear load conditions is shown in Fig. 10.15. The proposed SOSMO based SOSMC design also obtains a good unbalance compensation capability. The unbalance factor (UF) of the three-phase system currents is compensated from 26.8 to 1.9%.

The experimental result of the TC*LC*-HAPF under the start-up process is demonstrated in Fig. 10.16. When the TC*LC*-HAPF is switched on to the system grid, its DC-link voltage gradually increases to its reference value (60 V) within 2 fundamental cycles (40 ms). After that, the DC-link voltage is stabilized at its reference value (60 V). Figure 10.16 verifies that the proposed SOSMO based SOSMC design can perform a satisfactory DC-link voltage control performance including the start-up process.

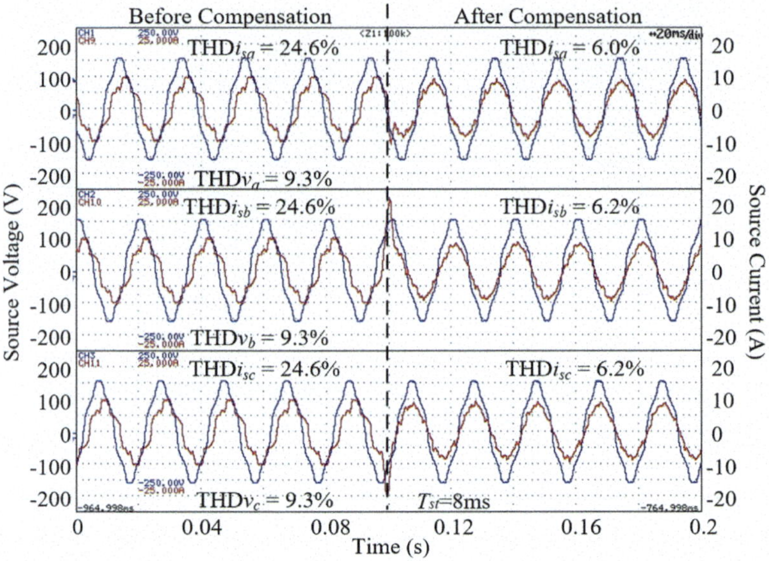

Fig. 10.14 Experimental results of TC*LC*-HAPF with proposed SOSMO based SOSMC under distorted grid voltage condition

10.8 Summary

An SOSMO based SOSMC design for TC*LC*-HAPF is proposed in this chapter, which ensures the TC*LC*-HAPF low steady-state error, fast dynamic response, fixed switching frequency, and good system robustness under parameter variation and non-ideal grid conditions. The SOSMO that provides a real-time observation of TC*LC* part's filtering capacitor voltage is firstly proposed, and its parameter design of the SOSMO under system parameter variation and non-ideal grid conditions are provided. Then, based on the proposed SOSMO, the SOSMC is proposed with its parameter design under system parameter variation and non-ideal grid conditions. Simulation and experimental results on a three-phase TC*LC*-HAPF are given, which verify the superior steady-state and dynamic performance of the proposed SOSMO based SOSMC design compared with the PIC, HCC, and MQPRC. Moreover, the system robustness under parameter variation, grid frequency variation and distorted grid voltage conditions, unbalance compensation capability and the DC-link control performance of the proposed SOSMO based SOSMC design are also verified by simulation and experimental results.

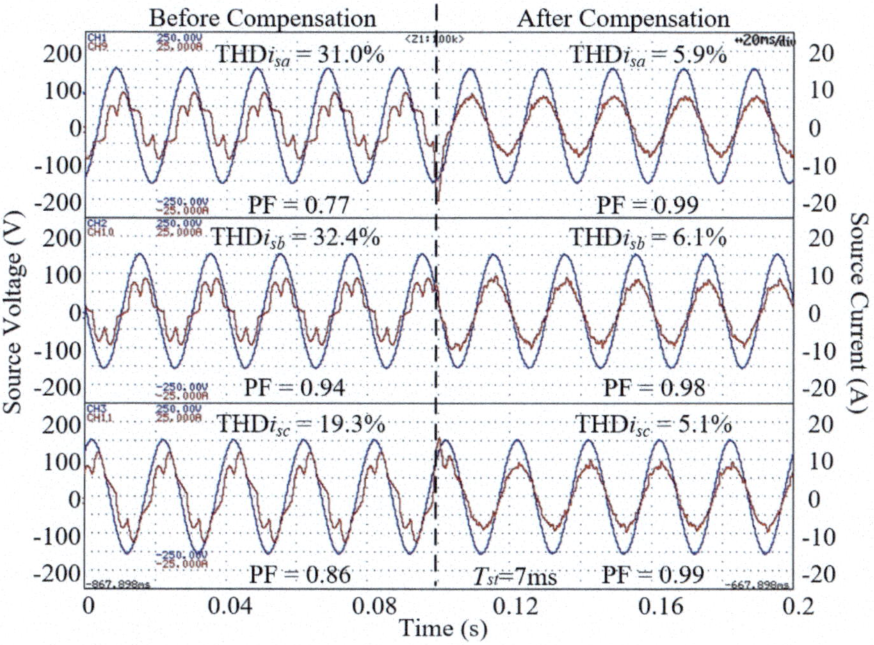

(a)

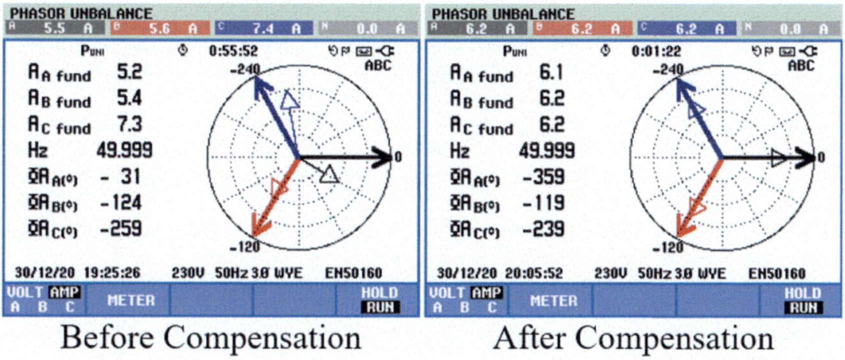

(b)

Fig. 10.15 Experimental results of TC*LC*-HAPF with proposed SOSMO based SOSMC under unbalance nonlinear load condition. **a** load voltage and system current waveforms. **b** load voltage and system current phasors

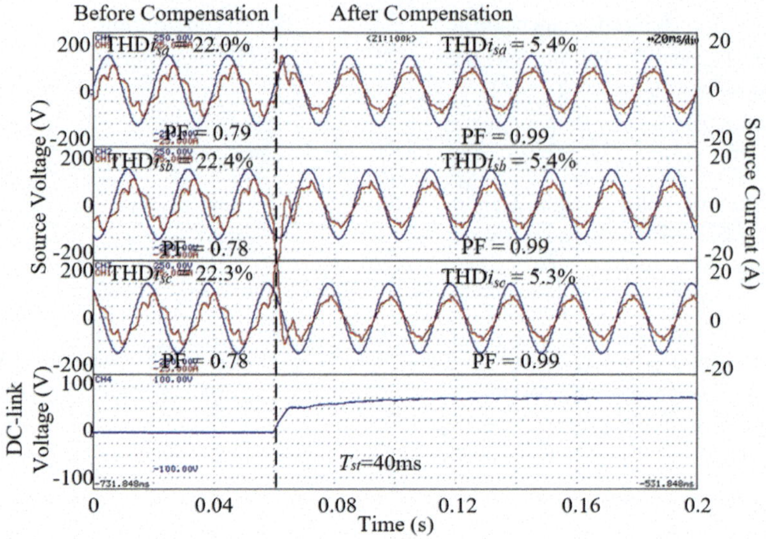

Fig. 10.16 Experimental results of TCLC-HAPF with proposed SOSMO based SOSMC under start-up process

References

1. S. Rahmani, A. Hamadi, K. Al-Haddad, L.A. Dessaint, A combination of shunt hybrid power filter and thyristor-controlled reactor for power quality. IEEE Trans. Ind. Electron. **61**(5), 2152–2164 (2014)
2. IEEE recommended practice for emergency and standby power systems for industrial and commercial applications. IEEE Std. 446-1995 Orange Book (July 1996), pp. 1–320
3. IEEE Standard 519-2014, Recommended practices and requirements for harmonic control in electric power systems (N. Y. IEEE, 2014)
4. R. Seeber, M. Horn, Stability proof for a well-established super-twisting parameter setting. Automatica **84**, 241–243 (2017)
5. J. Stoer, R. Bulirsch, *Introduction to Numerical Analysis* (Springer Science & Business Media, 2013)
6. S. Koch, M. Reichhartinger, Discrete-time equivalents of the super twisting algorithm. Automatica **107**, 190–199 (2019)
7. TMS320F281x Digital Signal Processors datasheet. https://www.ti.com/product/TMS320F2812
8. H. Akagi, E.H. Watanabe, M. Aredes, *Instantaneous Power Theory and Applications to Power Conditioning* (Wiley, 2017)
9. D. Yazdani, M. Mojiri, A. Bakhshai, G. Joós, A fast and accurate synchronization technique for extraction of symmetrical components. IEEE Trans. Power Electron. **24**(3), 674–684 (2009)
10. V.I. Utkin, *Sliding Modes in Control and Optimization* (Springer Science & Business Media, 2013)
11. L. Wang, C.-S. Lam, M.-C. Wong, A hybrid-STATCOM with wide compensation range and low DC-link voltage. IEEE Trans. Ind. Electron. **63**(6), 3333–3343 (2016)
12. L. Wang, C.-S. Lam, M.-C. Wong, Unbalanced control strategy for a thyristor-controlled LC-coupling hybrid active power filter in three-phase three-wire systems. IEEE Trans. Power Electron. **32**(2), 1056–1069 (2017)

13. C.-S. Lam, L. Wang, S.-I. Ho, M.-C. Wong, Adaptive thyristor-controlled LC-hybrid active power filter for reactive power and current harmonics compensation with switching loss reduction. IEEE Trans. Power Electron. **32**(10), 7577–7590 (2017)
14. L. Wang, C.-S. Lam, M.-C. Wong, Selective compensation of distortion, unbalanced and reactive power of a thyristor-controlled LC-coupling hybrid active power filter (TCLC-HAPF). IEEE Trans. Power Electron. **32**(12), 9065–9077 (2017)
15. L. Wang, C.-S. Lam, M.-C. Wong, Analysis, control, and design of a hybrid grid-connected inverter for renewable energy generation with power quality conditioning. IEEE Trans. Power Electron. **33**(8), 6755–6768 (2018)
16. C.-S. Lam, M.-C. Wong, Y.-D. Han, Hysteresis current control of hybrid active power filters. IET Power Electron. **5**(7), 1175–1187 (2012)
17. W.-K. Sou, C.-W. Chao, C. Gong, C.-S. Lam, C.-K. Wong, Analysis, design and implementation of multi-quasi-proportional-resonant controller for thyristor-controlled LC-coupling hybrid active power filter (TCLC-HAPF). IEEE Trans. Ind. Electron. **69**(1), 29–40 (2022)
18. A. Levant, Sliding order and sliding accuracy in sliding mode control. Int. J. Control. **58**(6), 1247–1263 (1993)
19. A. Levant, Principles of 2-sliding mode design. Automatica **43**(4), 576–586 (2007)
20. V. Utkin, Discussion aspects of high-order sliding mode control. IEEE Trans. Autom. Control **61**(3), 829–833 (2015)

Chapter 11
Conclusions and Prospective for Further Work

11.1 Conclusions

This book introduces and discusses the advanced controller design of three-phase four-wire *LC*-HAPF and three-phase three-wire TC*LC*-HAPF to further improve their compensation performance. The key contributions of this book are summarized below:

(1) The DBCC design for the *LC*-HAPF that provides a cost-effective power quality compensation solution with small steady-state error, fast dynamic response and low output current ripples is proposed in Chap. 2. Its effectiveness and superior compensating performances are verified by both simulation and experimental results in comparison with other representative controllers.

(2) The design of improved MPC with signal correction technique for the *LC*-HAPF is proposed in Chap. 3, which can track the reference compensating current with fast dynamic response and low steady-state error. The compensation performance is verified by simulation and experimental results in comparison with the conventional MPC and other representative current controllers.

(3) The VPIC that achieves low steady-state error, fast transient response, fixed switching frequency, and good system robustness even under a lower DC-link operating voltage is proposed for the *LC*-HAPF in Chap. 4. The VPIC performance is analyzed theoretically in comparison with other representative controllers, and verified by simulation and experimental results.

(4) The SOSMC design for the *LC*-HAPF that keeps good steady-state and transient dynamic performance even under system parameter variation, grid frequency variation, and grid voltage distortion is proposed in Chap. 5. The superior compensation performance and system robustness are verified by both simulation and experimental results in comparison with other representative controllers.

C.-S. Lam et al., *Advanced Controllers Design for Hybrid Active Power Filters*,
https://doi.org/10.1007/978-981-97-6847-9_11

(5) The MQPRC with gain scheduling algorithm that improve the steady-state performance with lower steady-state error, lower output current ripple and noise under both inductive and capacitive loads is proposed for the TC*LC*-HAPF in Chap. 6. The parameter design and gain scheduling algorithm are also introduced. Simulations and experiments are conducted to verify the effectiveness and performances of the MQPRC with proposed gain scheduling algorithm.
(6) The HSS model for the TC*LC*-HAPF to address its nonlinearity issue is proposed in Chap. 7, which can keep both high accuracy and LTI property. Based on the HSS model, the HOC that achieving high steady-state accuracy, fast transient response and fixed switching frequency is proposed and verified by both simulation and experimental results in comparison with the state-of-the-art MQPRC.
(7) The real and simplified predictive control model for the TC*LC*-HAPF are introduced and proposed in Chap. 8. Based on the simplified predictive model, the FS-MPC with coupling capacitor voltage observer for the TC*LC*-HAPF is proposed, which can provide a fast transient response, low steady-state error and good robustness under both inductive and capacitive load conditions. The effectiveness and performance are verified by simulated and experimental results.
(8) The FS-HMPC which is a combination of FS-MPC and hysteresis SMC is proposed for the TC*LC*-HAPF in Chap. 9. It provides the lowest steady-state error and fastest transient response among the existing control strategy for the TC*LC*-HAPF. Also, the robustness is improved compared to the FS-MPC. Its performance is verified by simulation and experimental results.
(9) The SOSMO based SOSMC that provides high steady-state accuracy, fast transient response, fixed switching frequency, good system robustness, and grid disturbance rejection capability is proposed for the TC*LC*-HAPF in Chap. 10. Various simulation and experimental results are carried out in comparison with the existing TC*LC*-HAPF controllers to verify the control performances of the SOSMO based SOSMC.

11.2 Perspectives for Future Works

Based on the research conducted in this book, there are many issues that warrant further exploration in future studies:

(1) The existing controller design methods for the HAPFs depend on the prior knowledge of the system model and system parameters. Therefore, an attractive and possible research direction is applying some model-free controllers inspired by fuzzy logic and artificial intelligence (AI) methods on the HAPF topology.
(2) The current control study for HAPFs focus on the device level for the end user. The further study will be the optimal coordinate control study on the system level between different power quality compensators to handle the power quality issues in distribution networks.

(3) With the development of carbon neutral policies, power quality compensation together with renewable energy injection will be the future development of power grid. The corresponding topology and control strategy studies of the inverter are necessary.

The manufacturer's authorised representative in the EU is Springer Nature Customer Service Centre GmbH, Europaplatz 3, 69115 Heidelberg, Germany. If you have any concerns regarding our products, please contact ProductSafety@springernature.com

Printed and bound by CPI Group (UK) Ltd, Croydon, CR0 4YY
01/10/2026
02225969-0002